VALUE ANALYSIS TEAR-DOWN:

A New Process for
Product Development and Innovation

Yoshihiko Sato
and
J. Jerry Kaufman

Industrial Press Inc.
and
Society of Manufacturing Engineers

Library of Congress Cataloging-in-Publication Data
Sato, Yoshihiko.
   Value analysis tear-down: a new process for product development
and innovation/ Yoshihiko Sato and J. Jerry Kaufman.
      p. cm.
   Includes bibliographical references and index
   ISBN 0-8311-3203-5 (professional / textbook: alk. paper)
      1. Value analysis (Cost control) 2. Industrial productivity. 3. New
products. 4. Engineering economy. I. Kaufman, J. Jerry. II. Title.
HD47.3.S38 2005
658.5'75--dc22

                                             2004019919

Value Analysis Tear-Down
First Edition 2005

Industrial Press Inc.
200 Madison Avenue
New York, New York 10016

Sponsoring Editor: John Carleo
Text and Cover Design: Janet Romano

TABLE OF CONTENTS

iii

ACKNOWLEDGEMENTS

Merging one's thoughts with a co-author to form a consistent, seamless, meaningful text can be challenging. The difficulty of the writing task is compounded when the co-authors live in different countries, speak different languages, and have different cultures. That this dream of Sato-san and myself in writing this book became a reality is credited to two people thanked here for their significant assistance and guidance.

William Christopher, President, Publication Services; – whose editorial skills are superb. Bill's ability to cut through lengthy passages to capture the writer's intent is exceptional. As a focused sounding board Bill presents his ideas and suggestions with the patience of a diplomat. I can still hear his sage advice; "less is better."

Kiyoo Narasaki, Translator, Interpreter, former Liaison Staff to the Society for Value Engineering (SJVE) (retired), and dear friend; – who built a cultural bridge allowing the free flow of communications between Sato-san and myself. Narasaki-san has the unique ability to translate the subtle and flowering Japanese expressions to direct American English, and reverse the process. Narasaki-san's assistance was invaluable in managing the constant flow of writing traffic across the ocean.

Not to be forgotten, my wife Harriet, for her understanding and for knowing when to protect my isolation and when to break in and insist that I should occasionally "smell the flowers." To my partner and co-author in this venture, Yoshihiko Sato, for developing this elegant methodology and allowing me to incorporate the principles of value analysis, making VA Tear-Down a more powerful competitive industrial discipline.

FOREWORD

Ryuichi Seguchi, President, Hitachi Construction Machinery Co., Japan.

I am honored to have been asked to contribute a Foreword to this book by my two friends: Mr. Jerry Kaufman of Houston, Texas, and Mr. Yoshihiko Sato of Japan, both Fellows, of SAVE, and internationally known Value Management authorities. I particularly recall the event of the first European Value Management Conference, 1989, in Milan, Italy, where representing Japan's contributions, I offered a congratulatory speech, and Jerry and Sato-san respectively presented technical VM papers. It is, therefore, my great delight to see them collaborate on a meaningful new value methodology and to be allowed to join them in publishing this important book.

My own career started as one of the earliest VA practitioners in Japan, and I have always been positive in adopting and promoting the Value Analysis and Tear-Down disciplines within my own company. I appreciate and endorse VA's excellence based on its proven performance at Hitachi. As is widely known, most of Japan's major manufacturers and other corporations have effectively adopted VA/VE as one of their effective Value Improving Practices (VIP) management tools. Particularly at Isuzu, Mr. Sato has endeavored to nurture what he had learned from GM and develop it into a unique process that can be used not only in Japan, but also universally by industries across the world. Certainly, Sato's method has been and is playing an important role in contributing to the growth of Japanese manufacturing industry. His technique is now one of the essential benchmarking methodologies and is spreading to other market segments. The Tear-Down process stands on a mountain of successful projects that will see new and higher levels of achievement with the merger of Value Analysis with Sato's Tear-Down process.

I would never hesitate to enthusiastically endorse 1) this rare case of ambitious teamwork of such prestigious consultants as Jerry and Sato-san, and 2) their determination to propagate their highly productive ideas throughout the world. I am confident that, just like the original Japanese edition of Sato's Tear-Down book, VA Tear-Down will be adopted as an international standard for improving value.

Joseph F. Otero, Jr., CVS
Discipline Chief of Systems Value Management at Pratt & Whitney

This book brings discipline, rigor, and creative thinking to the tear-down process by combining it with the value methodology which is one of the most powerful creative problem-solving processes in wide use today. Tear-down is an important exercise for understanding advantages and disadvantages of a competitor's product. I would recommend this book to any engineering company seeking to catapult past its competition. It is written by two internationally recognized experts.

INTRODUCTION

The globalization of the industrial market has added many new competitors, giving consumers a wide choice of products and features, with higher quality and lower prices. Adding to this mix, accelerated technology and the customer's demands for "newer, better, cheaper" makes staying ahead of the competition a full-time, expensive investment. Not long ago, an electronic product's manufacturer could plan on a five-year product life before obsolescence and competition required new or improved models. Today a manufacturer begins planning a product replacement before the product is even fully developed, let alone introduced to the market.

Two tested and proven disciplines available to assist the manufacturer in keeping pace with competition are Value Analysis and Tear-Down. Value Analysis and Value Engineering address product improvement and development by focusing on alternate ways to achieve needed functions for less cost. Tear-Down is a process that dissects competitors' products to uncover, analyze, improve, and incorporate those competitive features that appeal to the customer's sense of value.

Both disciplines require a structured building-block analysis, discipline, creativity, and marketing principles, making the merger of Value Analysis with Tear-Down a well-suited marriage for improving a product's competitiveness. VA Tear-Down merges the disciplines of Value Analysis with Tear-Down, creating a highly cost- and performance-effective process to match today's marketing and technology dynamics.

That the tear-down principles should be adopted and improved by Japanese industry should bring no surprises. In the export industry, which has long been the mainstay of the Japanese economy, recent fluctuations in exchange rates, a declining economy, and significant improvements in overseas industry have cancelled out the competitive advantage long enjoyed by Japanese products. Japan's aggressive pursuit of quality and innovation have created a new level of best practices, one that has been adopted by all as universal industrial standards. The "bubble economy" is now a thing of the past. Sluggish domestic demand and low facility-use rates have made domestic competition even keener, and Japanese industry is struggling to regain its global initiative. Ironically, yen appreciation, if it comes again, will make their salaries the highest in the world.

Japanese companies are forced to pay high salaries while their profits are shrinking. Therefore, improvement activities are one of industry's major competitive challenges. The VA Tear-Down method introduced in this book is one of the improvement methodologies that have proven to be effective in improving competitive position. The word "tear-down" might not be familiar to some readers and may imply some difficult process. Yet, VA Tear-Down is a relatively simple process.

The chapters that follow will guide you through the principles and applications of VA Tear-Down. The techniques presented in the following chapters will lead to finding new ideas by using our senses in combination with our knowledge. The procedures are relatively simple. Find advantages, combine them, and create something new out of such a combination. This book introduces logic built on the basis of experience, explains the concept of the VA Tear-Down method, describes specific procedures, and gives examples of actual applications. The book is arranged so that readers can improve products and operations by applying the procedures as they are presented. As you go through the book, I am sure you will find the VA Tear-Down method an effective and practical way to improve products and procedures, add value, and reduce cost.

So much for the introduction. Why not start now?

CHAPTER 1

VA TEAR-DOWN: WHAT IT IS, HOW IT DEVELOPED

In "Wheels," a novel about developing and manufacturing automobiles, Arthur Hailey refers to a room where competitive products are disassembled, dissected, and evaluated against the manufacturer's product offering. U.S. automakers were using a form of product dissection and analysis long before Value Analysis (VA) Tear-Down was developed as a value analysis process to stimulate ideas for product improvement. The method learned from General Motors is similar to the process described in "Wheels." In that 30-minute GM demonstration, the lesson learned was that there is a systematic way to disassemble and compare competitors' products and production processes.

GM-type methods were being used in the daily operations of many manufacturing companies, with slight variations among companies and products. One popular process rooted in the method is "reverse engineering," which reverses the product development process. Reverse engineering starts with a competitive product's performance, dissects the product to relate the components' contributions to product performance, and in this way works back into the conceptual phase of the product's development. However, the methods used by automakers and consumer appliances manufacturers in the United States, Germany, England, Korea, Taiwan, and Japan was little more than disassembly, inspection, and simplistic comparative analysis. The process described in the chapters that follow is far beyond these methods.

DEFINITION

VA Tear-Down is a method of comparative analysis in which disassembled products, systems, components, and data are visually compared; and their functions determined, analyzed, and evaluated to improve the value adding characteristics of the project under study.

Descriptions of the definition elements are:

Comparative Analysis – comparing two or more elements of a product or system having the same function

Product – the end item sold to the market being served

System – an active part of a product consisting of multiple components

Component – a single part or multiple pieces forming a single replaceable part

Data – elements of information that, when combined, form the basis for analysis

Function - a description of an intended action upon a defined object necessary to achieve a desired purpose

Although this definition may appear simple, it incorporates the powers of observation, deduction, and a broad range of value analysis techniques. This new approach separates VA Tear-Down from the simple disassembly and comparison techniques of conventional teardown.

The VA Tear-Down method provides a means to do much more than compare what has been physically disassembled. Using value analysis techniques helps users understand problems and uncover improvement opportunities in a systematic, analytical way. Many engineers initially regarded the teardown process as a simple disassembly and inspection process. But VA Tear-Down also proved an effective way of performing reverse engineering, new product development, model additions and modifications, and competitive analysis studies.

The term *disassembled* in the above definition means dissecting the project being studied and scrutinizing the details of a product or process focusing on their function contribution. The data includes the analysis of the processes that produced the components in addition to the data obtained as a result of analyzing the components. The word *visually* in this definition stresses the need to compare a variety of alternate components that perform essentially the same value function and verify the selections.

In Japan, comparative analysis tends to stimulate creativity. Engineers are sensitive to differences. If they lag competition, they want to overtake. If they lead, they want to maintain that lead. "If you don't recognize this propensity of engineers, you are letting a great opportunity slip," says Hirosi Okukawa, the Manager of the Value Engineering Center, Cost Planning Department, at Mazda.

There are some differences even in identical twins. Even the human face is not symmetrical comparing its right and left sides. When shopping for an item, two products performing the same function may look similar, but if other companies produce the product there are many subtle differences. By scrutinizing such differences and analyzing the advantages and disadvantages of each feature or attribute we can conceivably come up with a synergistic product that is better than the two being compared. In the VA Tear-Down process, users determine the product's functional advantages, the cost to perform those functions, and whether the features or attributes contribute to the value of the product as determined in the market place.

In addition to benchmarking whatever advantages there are in the products being compared, the objective of VA Tear-Down is to stimulate new ideas for function improvement and cost reduction. Human nature does not easily translate written information or knowledge into specific ideas. Examining an actual product brings all senses into play. What one sees, feels, hears, or otherwise perceives stimulates one's thought to create new and better ideas through the methods of VA Tear-Down.

FROM CONCEPT TO APPLICATION

The chronology that follows describes the history of the VA Tear-Down method, tracing its development and widespread acceptance, in order to provide a general understanding of the principles and their application.

In the early 1970s Yoshihiko Sato, the co-author of this book, attended a comparative analysis teardown demonstration by General Motor's consultants to Isuzu. That demonstration inspired the development of a process in which competitors' products were disassembled, displayed, and component improvement ideas developed, by value analysis methods.

The contributions made by VA were to focus on the function of the product and its components under study. Value Analysis is the study of functions and their relationship to other dependent functions. A physical part being studied in teardown is the result of some designer's concept of the way a function, or group of functions, should be implemented. Value analysis differs from other cost improvement initiatives by questioning, defining, and analyzing the functions of components, rather than determining how a "part" could be made better, or cheaper. Understanding the function being served by the component can significantly alter the component's geometry, integrate its function with other components, or eliminate the need for the function and its part.

By the close of the 1970s, the changes made using VA Tear-Down at Isuzu translated into significant cost improvements while also improving the quality and function of the product. On the basis of those initial successes, a library of case studies was collected, analyzed, and developed into a methodology.

The VA Tear-Down process was formally introduced and later integrated into the manufacturing technology at Isuzu Motors. A large VA laboratory was created. The laboratory contained dedicated teardown rooms containing competitor's disasembled products and the products of some fifty parts suppliers, each categorized by function, component, sub-system, system, and product. The design and production staffs at Isuzu use the VA laboratory for Value Engineering and other product improvement activities. As the VA Tear-Down process became institutionalized at Isuzu, the process and its history of successes were shared with other companies.

VA -related techniques in the early 1980s were developed at Isuzu. These techniques include the weight unit-cost method (relating cost to weight) and the VT (Value Target) method, or designing to a target production cost, thereby making product cost an engineering requirement. A technical paper by Yoshihiko Sato was published and presented at a National Convention on Standardization. GVE (Group Value Engineering method), a VA pilot program, was launched at Mazda.

By the mid 1990s, Mr. Sato attended the annual conference of the Society of American Value Engineers (SAVE) and the first European Community Value Engineering Conference, among other industrial conferences. In each, he represented Japan's manufacturing sector and presented papers on VA Tear-Down.

By the end of the 1990's Mr. Sato conducted numerous VA Tear-Down assignments in South Korea and Taiwan. He received three commendation awards from the Taiwan Industrial Foundation.

Since its introduction, Mr. Sato has presented the VA Tear-Down method at the Daily Industrial News seminars held across Japan, and at many academic conferences and institutional seminars. He has worked to expand VA Tear-Down principles and applications to make the concept applicable to a broad range of industries and products.

Currently, VA Tear-Down continues to gain recognition and acceptance in the manufacturing sector of Pacific Rim manufacturing companies as more companies adopt and apply the VA Tear-Down process in product development and improvement activities.

THE THINKING PROCESS

VA Tear-Down combines the traditional "dissect and analyze" methods with Value Analysis to stimulate the search for new and better ideas.

How does the VA Tear-Down method produce new ideas? What elements in the process are the triggers that stimulate the thinking process? Thinking, the ability to reason is a human attribute. In their study of creative behavior, psychologists note that humans recognize things and take actions on the basis of the functions of the brain, as shown in Figure 1-1. Two or more of these functions (imitation, research, cooperation) work in harmony when an idea is conceived. Imitation and research are two functions that play a major role in developing ideas.

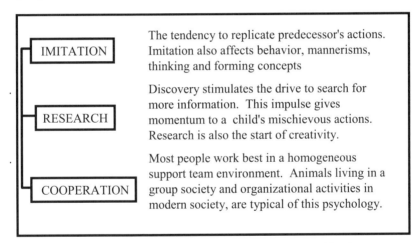

Figure 1-1 Three Functions of the Human Brain

Unlike in lower forms of animals, the human brain has a huge storage capacity for knowledge. The human cerebrum consists of the occipital lobe, parietal lobe, frontal lobe, brain stem, and two temporal lobes. Knowledge is stored in the temporal lobes. The entire brain, weighing about 1.2 to 1.4 kg, contains as much as 14 billion cells and its surface skin covers as much as 22,000 cm2 in area.

The size and weight of the brain is relatively constant in mass regardless of the size of the head. It is encouraging that we each have the same brain size as Nobel Prize winners. The amount of memory stored in the right and left temporal lobes does not differ much from one person to another either, which means that all humans can process about the same amount of information or knowledge stored in his or her brain. There are, however, some factors that do make a difference in how that knowledge is received and retained. These include experience, degree of interest, amount of learning, and timing at which the information is processed, and other experiences. Other significant factors are the way and how often the stored memory is used and how imitation or research biases the thinking process. Those studying the creative process agree that creative thinking uses the right side of the brain, while judgment is a left-brain function.

Once acquired, knowledge is stored in the temporal lobes. When some information or stimulus (often referred to as a language signal) is received, the signal goes into the temporal lobes through an organ called the hippocampus. The information or knowledge stored in those lobes then moves to the frontal lobe where it is combined or processed for use.

Language signals are received through the eyes, ears, or other sensory organs. Humans differ from animals in their ability to stimulate their own brain without external stimuli. When a person thinks of anything of interest, the stimuli turn into language signals, which in turn trigger whatever is stored in the temporal lobes. New language signals are then produced, which help retrieve the memory.

Smart, clever, and creative people possess the ability to gather, store, and retrieve data in a way that combines and sends proper signals from the temporal lobes to the frontal lobe in a timely manner for proper processing. The result of such processing is an ability to analyze, conclude, express, and transfer knowledge into VA Tear-Down language signals. Applying the technologies of Value Analysis then produces successful projects for both improvement and innovation.

INTEGRATING VA WITH TEAR-DOWN

The thinking process starts with the stimulus of the language signal. When you read this book, your imagination or thinking function is focused on the written text and illustrations. You are not (or should not be) thinking about politics or sports, which are topics not related to the subject of this book. Because no outside language signals are triggered that relate to politics or sports, the reader maintains focus on the subject. However, the moment you see these words, some readers may think of a recent political event, or of their favorite professional baseball player. This is also the result of the language signal at work.

Some triggers in VA Tear-Down that stimulate language signals are:

1. *Comparative method:* Comparing the description and illustrations of like products for the purpose of selecting the best features

2. *Actual products and data:*Physically dissecting competing products and evaluating their component characteristics

3. *Choose and combine advantages:* Same as 2 above, but using the comparison analysis to create a new option

4. *Actual products data and functions:* Same as 2 and 3 above, except determining the functions of the components, selecting the best functions, then comparing competitive approaches that best achieve the desired functions; combines Value Analysis with conventional dissection practices, and is key to the VA Tear-Down process

As mentioned earlier, the VA Tear-Down method is a comparative analysis process in which ideas are created by analyzing competing products. The spring of ideas, that is, analyzing the competitor's products used in the comparison, creates language signals. This is a form of imitation and creation technology.

Figure 1-2 shows several containers that are used for different purposes, but they have common functions of "storing liquid," "pouring liquid," and "carrying liquid." In addition, some have the features of spraying water or keeping water temperature constant. In value analysis, this is referred to as "function definition," but the attributes and the shapes that enable each pitcher to achieve its objectives are not the same.

There are two basic approaches used in the comparative analysis process. The first is to look at the products, compare them, and then select. For example, the differences in pitcher handles were each designed to suit a particular need. In this simple, imitation-type thinking, we would select the shape that best suits our purpose.

The second VA Tear-Down approach is to let the sight of the pitchers turn itself into the language signals that help us select and regroup information and ideas from our memory. This is a research-type thinking process stimulat-

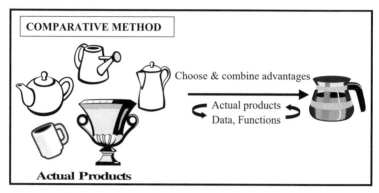

Figure 1-2 An Example of VA Tear-Down

ed by imitation. The result is a newly configured handle, different from those being analyzed.

The handle of the improved pitcher shown in Figure 1-2 is the product of many ideas prompted by the language signals that were sent out by the handle of the earthen teapot. The ideas include those concerning materials, manufacturing processes, and surface treatment.

There is, however, a limit to what one person can think of simply because creative thought is limited to what is stored in the temporal lobes. No one can trigger memories that have not been experienced. This is the reason why multi-functional teams are used in VA Tear-Down projects. Simply stated, "Two heads are better than one."

The practice of VA Tear-Down begins with simple imitation. As confidence and proficiency in the process develops, more creative approaches and innovative solutions emerge as a result of triggering more language signals

VALUE ANALYSIS' CONTRIBUTIONS
TO THE TEAR-DOWN PROCESS

Value Analysis was formally created in 1945 by L. D. Miles at General Electric and was introduced into Japan in 1955. It has been widely used in the United States and other countries including Japan as a method for cost reduction and product improvement through function analysis.

In the mid 1950s Japan's consumer products manufacturers made dramatic progress gaining market share from their global competitors in some targeted fields. Major credit for this success is attributed to Japan's dedicated use of these product and process control techniques. The other two are quality control (QC) and industrial engineering (IE). In Japan, these three initially different techniques have merged with Japan's work ethic and grown into a comprehensive manufacturing technology. This combination has enabled Japan to win market share over U.S. and European companies in many markets, offering functionally-responsive, cost-effective, quality products and services to an informed market.

In a SAVE International Conference in the United States, a speaker observed, "Japan has taken Miles and Deming (the founder of QC) away from the United States." Many initiatives have branched out of Miles' original value analysis concepts. Among these initiatives are quality function deployment (QFD), failure tree analysis (FTA), and failure mode and effect analysis (FMEA). Value Analysis and Value Engineering (the term to describe product development rather than product improvement) are now among the most widely-used methods in Japan for controlling and directing technology. The primary emphasis of value analysis is on understanding, identifying, and classifying product-related functions. Once defined, the creative value engineering steps encourage a broad search for ideas to implement those functions and produce innovative products.

In today's market, it is essential that our products have some differences in the way functions are implemented, as well as the cost to produce, in order to separate us from our competitors and attract the attention of potential customers. Although it is taken for granted that our products have functions that our customers are willing to pay for, the proliferation of product offerings and model options have made determining what functions and features the customer wants a difficult marketing search. What functions in products, systems, or processes would raise the value-adding perception of our customers, including aftermarket systems and services? In short, how can we improve function and reduce cost? Value analysis is a discipline to deal with such issues; it is a very useful method for winning in the market place.

VA TEAR-DOWN AND CREATIVITY

As we saw in its definition, the VA Tear-Down method is rooted in value analysis.

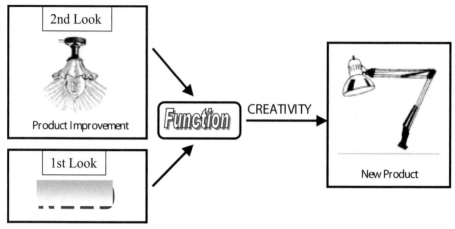

Figure 1-3   Applied Creativity—VA Methods

Defining the functions is a process characteristic of value analysis methods. VA is a process where the components of an existing system or product are "blasted" into pieces, which are then translated into the more abstract concept of functions. Determining the use of the individual parts and how they contribute to the system are translated into two-word, verb-noun descriptions (language signals) that define the functions of the parts. Addressing the function of the part, rather then how it is used, will open the search for more creative options. This will be explained in Chapter 2. Because the creative process focuses on unique ways of performing functions, rather then focusing on the actual product or its components, our minds are encouraged to stretch our imagination in searching for new approaches.

Conventional teardown methods do not define functions. Competing products are visually compared to find their features and advantages. The first step is to find differences, and then take advantage of the best approach by incorporating it into our own products. Conventional tear-down triggers the language signals by observing the physical product or data, while value analysis triggers language signals by identifying the functions performed by the components under study, or blasting the product into basic function terms. Once a needed function is identified in the development phase, many concepts can be evaluated to find the best way of performing those functions. Conventional tear-down methods need the physical product to dissect. VA Tear-Down starts from "Blast", goes to "Create" and then to "Refine." Conventional tear-down has no "Blast" (the analysis of function) step. VA

Tear-Down begins with function determination.

Chapter 2 describes the relationship between Value Analysis and VA Tear-Down in more detail.

IMPROVEMENT METHODS IN VA TEAR-DOWN

On the basis of thinking principles, the value engineering method can be referred to as a creative thinking process (research type), and the teardown process can be referred to as the comparative method (imitation type). Figure 1-4 classifies improvement techniques on the basis of thinking principles. The Mona Lisa method and the weight per unit-cost method are explained later in the book.

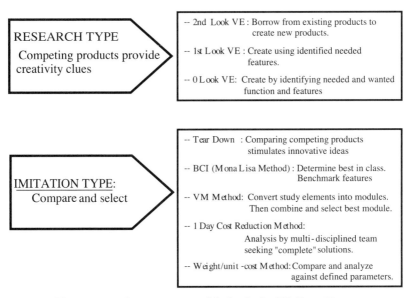

Figure 1-4 Improvement Methods in VA Tear-Down

VA TEAR-DOWN AND IMPROVEMENT/INNOVATION

We are always improving existing things—products, processes, systems, etc.—and creating new things. A company dies when these activities stop. Companies must continuously improve and innovate to maintain market position. How are these two concepts different? Improvement means modifying existing things to achieve incremental enhancements. Small, but continuous improvements, piled one on top of another, are essential to the business success of any corporation. As a Japanese proverb says, "Little and often make a leap in time." Innovation means making major change in concept and configuration. VA Tear-Down produces both improvements and innovations.

Improvement and innovation cause different degrees of change. With this distinction, VE is innovation focused, while VA is more intended for improvement. Taking the best parts of both processes, the VA Tear-Down method integrates Value Analysis and the principles of Value Engineering with the creative stimuli they produce. The synergistic effect is the emergence of a powerful technique for product and process improvement and innovation.

In our technology-accelerating, competition-oriented society, improvement alone cannot provide the strategic advantage to win over competitors. Some level of innovation is also needed to gain strategic advantage.

The VA Tear-Down method described in this book, producing both improvements and innovations, can help achieve a high level of success, with a minimum of investment and risk.

The VA Tear-Down process is a structured building process that guides creative application. It begins with exercising the right side, or creative part of the brain and concludes with the left, or judgmental side. The process combines employees representing marketing, engineering, and manufacturing in a team environment, focusing on a well-defined product innovation or improvement objective.

The investment in capital and time in VA Tear-Down activities is significantly less than that needed for traditional applied research and development activities addressing the same business improvement objectives.

THE VA TEAR-DOWN METHOD AND ITS COMPONENTS

Conventional teardown comparative methods are limited to displaying what is to be evaluated, then letting the viewer's imagination and experience find differences. Lacking a systematic, structured approach, ideas are randomly produced without a focused objective or an organized way of processing them. Today's procedures, developed from past experience and the incorporation of the VA discipline, comprise a systematic way of comparison and analysis.

Any product includes a variety of factors measured in terms of cost or time. Some of these factors are: materials, construction, assembly, test methods, and those investment expenses classified as fixed and capital costs. The VA Tear-Down process focuses on each of these factors, comparing and analyzing them in search of problems or opportunities. Then the process sets about improving each item that has surfaced in the course of the analysis. Figure 1-5 illustrates the structure of VA Tear-Down.

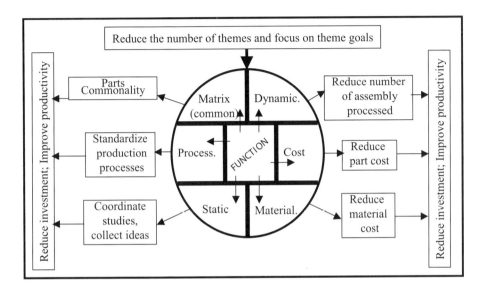

Figure 1-5 Structure of VA Tear-Down

The following are the VA Tear-Down method's major features:

The VA Tear-Down process makes it possible to concentrate on a particular area and objective called a *theme*, or an issue of concern.

Since a great number of ideas develop in the course of analysis, the analysis process can be directed to meet specific objectives or themes such as cost reduction, process reduction, commonality, introducing new functions, and improving existing functions. Suggestions for proposed solutions can also be collected and directed to a specific theme.

Many departments can simultaneously perform their analysis on a single project or product. The process does not require concentrating on the issues of a single function.

Static VA procedures (discussed later) enable the participants to share problems and expectations and discuss their resolutions to avoid resolution conflicts while focusing on their own projects. This allows the solutions of one department to emerge without creating bigger problems for another department.

VA Tear-Down can be used to resolve a specific approach or a specific objective such as reducing manufacturing process cost or operating cost.

VA Tear-Down should only be applied to selected essential themes.

VA TEAR-DOWN PROCEDURES

The six types, or applications of the process shown in the circle surrounding Function in the center of Figure 1-5 can be arranged sequentially, as shown in Figure 1-6. The illustration shows five analytical and one display application process. Each of the applications and its corresponding VA Tear-Down processes will be discussed later in detail. Any one or more, sometimes all, of the processes can be used to improve the value-added characteristics of a project.

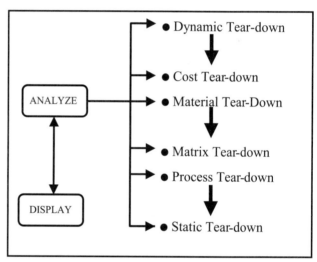

Figure 1-6
VA Tear-Down
Procedures

VA TEAR-DOWN APPLIED TO PRODUCT MANUFACTURING

The purpose of the VA Tear-Down process is to analyze and understand the competitive advantages of our own and competing products. The process also allows the products being torn down and analyzed to be benchmarked and to create product improvement goals. As a creative stimulus, the process encourages developing new ideas that will improve functions, features, and attributes of products in direct competition to a targeted competitor. The products are then displayed (static) and used for information sharing and networking with everyone concerned, including the top management, joint ventures, and major systems suppliers. Using the static displays to reinforce the business case presentations adds credibility and interest to the request for new or major product improvement investments.

This section discusses the effect of VA Tear-Down on manufacturing, specifically for producing component parts, and assembly testing of consumer products.

ANALYSIS – USING IDEAS

Validating and implementing the ideas collected through the VA Tear-Down process on existing products can produce some immediate, direct improvements such as lower material and direct labor cost, or improved functions. Product improvement ideas that cannot be immediately implemented are recorded and set aside for future use. Such ideas are stored in a "Product Improvement Library," or a product development idea bank. This database, properly used, can reduce product development lead-time and capital investments, as well as the unit cost. The ideas are also used to build business cases and justify the need for acquiring new capital equipment and advanced manufacturing technology.

SETTING TARGETS

The VA Tear-Down process can be used to focus on which parts need to be improved, and to what extent and in what sequence the company can achieve a competitive edge. This information is used to establish cost and performance targets for product improvement.

It isn't necessary to target every component in a product. Using Pareto's maldistribution rule, approximately 80 percent of a product's cost resides in 20 percent of its components. Identifying and addressing that 20 percent of the components makes the target cost program more manageable in planning necessary actions and establishing priorities for such actions. One should not establish product improvement targets solely based on financial considerations. Product improvement targets should also consider the needs and wants of the customer as well as the competitive environment in which the improved product is introduced. These factors, when combined with the financial objectives to justify product improvement investments, make the venture more credible and generate more enthusiasm than those targets established through some abstract financial business planning process.

DISPLAY – COORDINATION

By presenting the results of various analyses in an easy-to-understand physical display, concerned team members can use all their senses to easily recognize the competitive features or disadvantages as well as scope the problems that are inherent in their products. Thus, everyone participating in the VA Tear-Down activities can share their information about their individual concerns, ideas, and constraints, then agree to a corrective course of action. Such actions will affect everyone involved including design, development, production, and business management.

RECOGNIZING PARTS SUPPLIERS' COMPETITIVE CAPABILITIES

Few companies today design, produce, and assemble all of their product systems and components. It is a common practice that assemblers and parts suppliers cooperate to make a final product. As the Japanese yen appreciated and Asian manufacturers caught up with Japanese counterparts in product quality and functions, it became no longer necessary to restrict parts sourcing to a limited number of qualified domestic suppliers. Limiting supply sources could significantly detract from competitive cost and price advantages.

1. Determining the competitive capability of a finished product, function, and quality advantage can easily be determined, but analyzing a competitor's manufacturing costs are much more difficult. Knowing the competitor's price does not give a clue to its product cost, unless its price strategy, discount policies, internal accounting structure, and level of manufacturing technology are known.

2. The VA Tear-Down display makes it easier to determine what the competitor's product "should cost" by applying reverse engineering techniques to the dissected model. Such analysis will uncover information about the competitor's processes, assembly, materials, labor hours, capital equipment, and many other manufacturing details. VA Tear-Down is also used to evaluate parts supplier candidates.

3. Through the VA Tear-Down process, a supplier's development and technological capabilities and their competitive advantage contributions can be evaluated. Based on this evaluation, supplier selection as well as potential future partners can be decided. Information retrieved in support of such decisions includes current supplier capabilities, shortfalls, and investments needed to make those suppliers better partners.

COLLECTING NEW SUGGESTIONS FOR IMPROVEMENT

A company's technical staff normally performs VA Tear-Down analysis. However, this dramatic visual aid is used as a stimulus for other interested people. Those outside the product development loop, having different paradigms, offer fresh, good ideas that have been conceived from a different perspective. What would a plastics expert, or other material and process specialists, think when looking at a machined part? Entirely new ideas, not only about material but also manufacturing procedures, may emerge. Such spontaneous comments can be encouraged by creating an open-access display room in the procurement lobby to be viewed by suppliers and others waiting to meet with buyers. Supplier ideas are encouraged and collected by the technical staff and evaluated at a later date as illustrated below in figure 1-7.

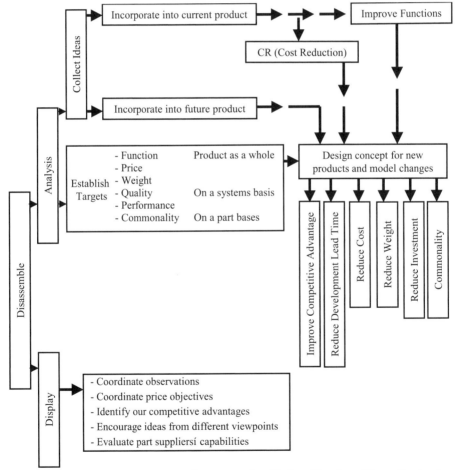

Figure 1-7 Effects of VA Tear-Down as Applied to Product Manufacturing

VA TEAR-DOWN APPLIED TO PARTS MANUFACTURING

End product producers place their name and reputation at risk for liability-related problems arising from the use of their products. Those who are most directly involved with mitigating liability risk represent the most enthusiastic users of the VA Tear-Down process. In Japan, consumer goods manufacturers such as automobiles and home electronic appliances have embraced and integrated VA Tear-Down as part of their operating culture. But the use of VA Tear-Down in parts manufacturing still lacks total acceptance and endorsement.

Most product producers outsource their components' manufacture to focus on assembly and test. They are turning more of the parts manufacture to more cost-effective supplier experts. This has encouraged the global growth of component manufacturers, which has increased the competitive pressures among those producers. That competition may even be sharper than that among end product producers. The VA Tear-Down method helps parts manufacturers

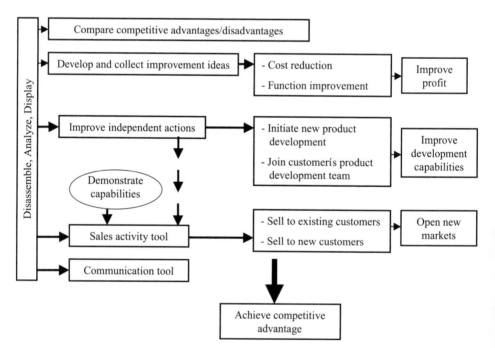

Figure 1-8 Effects of VA Tear-Down for Parts Manufacturers

improve their competitive position. In addition, the process can be used by part manufacturers as a marketing and sales tool by displaying how their parts can be integrated into the assembled product to improve cost and functionality. Figure 1-8 is a summary of VA Tear-Down for parts manufacturers.

There are few opportunities for component manufacturers to demonstrate their parts development capabilities. The VA Tear-Down display can be used to allow their end-product customers to use all the senses to visualize and physically examine existing parts and those under development. Such a sales aid not only strengthens ties with current customers, but also provides an excellent marketing tool for attracting new customers. A number of business cases collected by the authors over the last 15 years attest to the success of such practices using VA Tear-Down.

CHAPTER

2

VALUE ANALYSIS AND VA TEAR-DOWN

INTRODUCTION

The need for Value Analysis and VA Tear-Down is more obvious when product sales are in trouble than when things are going well. Well by what standards? The competition? Without our changing anything, we could go from good to poor by the competition changing the base line. Are we truly responsive to customer's need? Do we have product features and functions the customer pays for but doesn't need? Does the customer need features and functions we don't furnish? Have we investigated a range of alternative ways to satisfy the market requirements? From that range, have we selected the option that is the most cost effective, or are we producing "pet" ideas? Do we know which of our product features are value adding? Do we know the cost they incur or are those costs all buried in a lump-sum total? If there is room for improvement, it is better for us to find it before the competition does.

Value Analysis is fundamentally different from traditional approaches to design, cost reduction, industrial engineering, and production engineering. The key difference is that Value Analysis offers a method for identifying and selecting the lowest-cost approach, from many alternative methods, to satisfy the proper function needs. These features make Value Analysis the ideal methodology to combine with tear-down to create a responsive solution to the above questions. That solution is called *VA Tear-Down*.

A single idea for reducing cost to meet a design requirement is not Value Analysis. Although the idea probably represents better value, there was no attempt to determine whether the idea represented the best value from a selection of alternatives, or whether the design requirements being satisfied represent the real problem. Value Analysis adds these dimensions to good engineering.

The benefits of using VA Tear-Down include: contributions to the goals of the profit plan and sales plan, the development and building of teams to problem solve, the application of creative thinking in daily job performance, and the development of a sharper sense of value as defined by the customer.

IN THE BEGINNING

The search for improved value that evolved into Value Analysis began during World War II. The General Electric Company, concerned with the difficulties in obtaining critical listed materials to produce war materials, assigned electrical engineer Lawrence D. Miles to the purchasing department. His mission was to find adequate material and component substitutes for critical listed material to manufacture needed war equipment. In his search, Miles found that each material has unique properties that could enhance the product if the design was modified to take advantage of those properties. Miles discovered that he could meet or improve product performance and reduce its production cost by understanding and addressing the intended function of the product.

Miles' approach to improving the value of existing products was to "blast, create, and refine." The "blast" step is accomplished in VA Tear-Down by dissecting our products and competitors products that have competitive advantages over our product. The "create" step is the detailed analysis of the disassembled products, identifying those functions of concern, and soliciting ideas for improving our product to regain and overtake the competitor's advantage. The "refine" step is selecting the most value adding, cost-effective ideas and preparing a business case for the implementation of the proposals.

FOCUS ON FUNCTION

Miles separated function (or the intent of the design expressed as "what it must do") from the characteristics of the design (or "how it does it"). Following the close of World War II, the country's manufacturing sector turned to design and produce consumer products. Miles applied the lessons learned to create competitively-produced consumer products for the GE Corporation. Focusing on function as the way to improve value, he expressed this discipline in the following relationship.

$$\text{Value} = \frac{\text{Function}}{\text{Cost}}$$

VALUE EQUATION #1

This seemingly simple relationship is the cornerstone of Value Analysis (VA). Although expressed simply, the relationship of function to cost has broad implications.

Expanding function, the principle value elements used in VA studies are classified as:

> Esteem Value or "Want"
> Exchange Value or "Worth"
> Utility Value or "Need"

Each decision to acquire goods or services includes one or a combination of all the value elements, where the sum of the elements results in a buy decision. Esteem value or "want" expresses the desire to own for the sake of ownership. Collectibles fall into this category, but more important to VA, the reputation of the company can carry a quantifiable level of esteem value. Well-known companies that earn the reputation or perception for product quality, after-market support, and innovation can command a higher price for essentially the same product and functions produced by lesser-known companies.

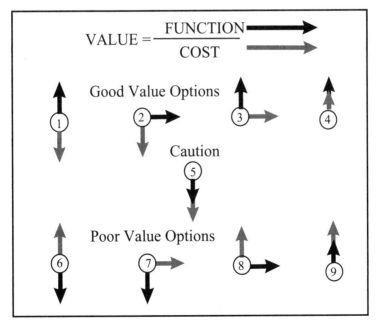

Figure 2-1 Function/Cost Relationship

Exchange value or "worth" describes the shift in value from the seller to the buyer. Improving the worth of an offering requires a good understanding of why the product interests the buyer, and how and when the buyer will use the product, and then incorporating those attributes in the design of the product.

Utility value or "need" is the primary value element addressed by the design engineer. Utility describes the performance and physical characteristics of the product, usually measured in engineering terms.

Function is defined as the intent or purpose of a system, product, or process operating in its normally prescribed manner. Using these defined values expands the value formula:

$$\text{Value} = \frac{\text{(Esteem) (Exchange) (Utility)}}{\text{Cost}}$$

VALUE EQUATION #2

Figure 2-1 (Function/Cost Relationship) illustrates the dynamics of the value expression and the value improvement solution paths.

The function/cost conditions in Figure 2-1 illustrate the various relationships between functions and cost that affect value. The four examples in the upper row show that improving the function can improve value. The lower row shows four examples of poor value. Reading across the top, from left to right:

1. Function Increase, Cost Decrease

Reducing cost does not necessarily result in a decrease in functions. The two factors are mutually exclusive. This illustration shows that creative engineering in product improvement can result in better functions at lower costs. It should be noted that reducing cost does not mean that we must reduce price. The customer pays for functions, not product cost.

2. No Change in Functions, Reduced Cost

This example illustrates the single focus on cost reduction without affecting the functions or features of the product. The target opportunity for this illustration is the factory process, inventory, purchased items, etc. This example assumes that reducing cost to produce the product, and subsequently, the price to the buyer will increase value.

3. No Change in Cost, Increased Functions

When adding functions as the approach to overcoming competitor advantage, it is critical to understand that value-adding functions are those that customers are willing to pay for. Adding functions and features that are not perceived as adding value may do more harm than good in trying to attract customers.

4. Increase Cost and Increase Functions

Again, only those functions that customers are willing to pay for should be incorporated in our products or substituted for less-valued functions. Improving functions without increasing perceived buyer value or worth will not increase market value. This condition illustrates that adding customer-wanted functions can offset price increases.

5. Reduce Functions and Reduce Cost

Reducing functions to reduce cost, as illustrated in the center of Figure 2-1, is a common way to achieve cost reduction goals. The box labeled "caution" identifies an opportunity to improve value by reducing or eliminating functions that are cost drivers. However, the value practitioner should not disturb those customer-sensitive functions that are the primary reason for the success of the product. Because customer-sensitive functions are not always obvious, the value practitioner should explore alternatives before deciding on this course of action to improve value. The danger is that we may be eliminating functions that accounted for the success of our product in the market place. Removing those functions will result in continued or accelerated drop in sales, regardless of the cost or price reductions. Many manufacturers do not have a clear idea of which functions offered in their products are highly valued by their customers. This is because we tend to take success for granted, but we spend much time and effort in determining why we failed. If manufacturers would take as much time and trouble to analyze why they succeeded as they do in analyzing their failures, they would better appreciate what makes customers value particular functions over other functions.

6 – 9. Poor Value Conditions

The conditions illustrated in items 6 through 9 are the reverse of the first four value-adding conditions. These conditions are poor value and should always be avoided.

The daily flower auction in Aalsmeer, Netherlands, is an example where all else being equal, price determines market value. The process is called a Dutch Auction. Buyers view flowers in lot quantities with a starting price. Every few seconds the large price clock lowers the price until a buyer signals a buy. The process continues until buyers purchase all the flowers. In this example, the buyer truly establishes the market value of the product.

WHAT'S IN A NAME?

As the search for cost reduction by analyzing functions grew into a procedure, Miles named this process "Value Analysis." The process evolved into a team activity directed at reducing high product and component costs while protecting the principle or basic functions of the project.

In 1945, the results of Value Analysis attracted the U. S. Navy's Bureau of Ships to commission L. D. Miles to train their personnel in the process. Miles recognized the difficult task of Navy personnel accepting the name "value analysis" because the Navy's table of organization had no analyst slot. On the advice of the Bureau, Miles selected the engineering department to be the home of the value process. To give organizational recognition to the process, the Navy renamed it "Value Engineering." Today, this renaming still causes much confusion.

Because industry and the Department of Defense use both names, Value Analysis and Value Engineering, new definitions emerged in an attempt to justify the terms by separating the process, which only led to more confusion. The Defense Department described Value Engineering as a "before-the-fact" activity, applying the value methodology during the product design phase. They defined Value Analysis as an "after-the-fact" activity, practicing the value process following design release, during the production of the product. These definitions follow the Japanese concept of VE/VA, on which VA Tear-Down is based. In Japan, Value Engineering is applied to the first look, or Stage 0, that is, the conceptual phase of product Development. Value Analysis is the name given to second look, or Stage 1, where product improvement of an existing product is the objective.

Because VA Tear-Down explores competitive advantages of existing products, the name "Value Analysis" and "VA Tear-Down" best describe the processes in this book. Whether called VE or VA, the methodology is exactly the same. Only the application differs.

Value Analysis is defined as:

An organized effort directed at analyzing the functions of goods and services to achieve those necessary functions and essential characteristics in the most profitable manner.

This definition also comfortably describes VA Tear-Down.

The following lists the key elements of this definition:

An organized effort – VA is a structured building block process that consists of defined steps called the "job plan." VA Tear-Down very closely follows the structure and sequence of VA.

Analyzing and achieving necessary functions – VA includes an organized effort to identify what the market furnishes and what the market needs, as opposed to perceived wants. The VA Process, as does VA Tear-Down, interfaces engineering and marketing to define the priority requirements from the point of view of the customer and includes the target price.

Essential characteristics – In addition to achieving the basic functions of the product or process, the product or process must also satisfy other requirements and attributes such as quality, time to market, safety, and maintainability.

In the most profitable manner – VA determines cost, generating and evaluating a range of alternatives that includes new concepts, reconfiguration, eliminating or combining items, and process or procedure changes. VA also considers the operation and maintenance of the product over its normal life expectancy to be the cost of ownership.

These elements interface marketing, engineering, and manufacturing. How does the term Value Analysis fit in? VA is a methodology. VA Tear-Down describes the application of this methodology.

THE JOB PLAN

Key to the VA methodology is the "Job Plan." The job plan is a disciplined approach consisting of sequenced steps that guide the VA team through the problem solving process. Miles developed the original five-step job plan that has withstood the test of time. Although users can modify the job plan to fit their unique requirements, the original form is common to every VA study and can vary in a number of ways. Common to all forms of the Job Plan are the following five steps:

1. Information

Evaluation of all available information relative to the VA project and translation of that information into function terms. In VA Tear-Down, the disassembly and analysis of competitor products is an elegant way to achieve this objective.

2. Speculation

Process of developing a large quantity of ideas (not solutions) that addres unique and creative ways to achieve those functions that relate to the problem definition. The VA Tear-Down display of the disassembled products, focusing on the issues of concern, stimulates visitors to exercise their creativity in suggesting improvement ideas.

3. Planning (or Analysis)

Evaluation of those ideas previously generated, using weighted guidelines, performance, and other requirements, to sift and sort for the best ideas. VA Tear-Down uses these techniques as described in later chapters.

4. Execution (or Evaluation)

Clustering of selected ideas into proposal scenarios and the evaluation of those scenarios including the financial, risk, and the implementation plans.

5. Reporting (or Presentation)

Preparation and presentation of recommended VA Tear-Down team proposals to a management board (or stakeholders), and seeking approval and funding to implement those actions to resolve the problem or develop the opportunity.

Although implementation is not considered part of the job plan, it is the most important step. Regardless of how effective the VA Tear-Down project may be, the value-adding and product-improvement results are only potential until the proposals are implemented.

Following the steps of the job plan in sequence requires discipline. This is a building-block process in which it is often necessary to repeat a step or two, but the value practitioner must never skip any steps. The VA Tear-Down team may go back to previous steps when the need for more information is warranted, but a step should not be skipped or taken out of sequence.

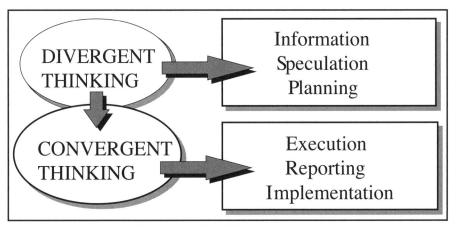

Figure 2-2 Job Plan Sequence

The job plan incorporates the creative divergent-convergent thinking process. Divergent thinking (sometimes called right-brain thinking) explores the unique by stimulating and applying the team's creative traits (see Figure 2-2). Convergent thinking (sometimes called left-brain thinking) explores relevancy, applying the principles of logic and quantitative analysis techniques. The information, speculation, and planning phases are considered divergent thinking. Execution, reporting, and the subsequent implementation action are considered convergent thinking.

The search to uncover the root cause problem is considered a divergent thinking process because it requires the application of creativity first to separate symptoms from root problems, and then to adequately define the problem in a way that leads to a resolution.

EXPANDING THE JOB PLAN

As the VA methodology progressed from reducing the cost of components to more complex issues, a "Pre-Event" step was added to the job plan process to address and sort through a greater volume of information and complex issues. The activities within the Pre-Event phase describe the planning and resolution of issues that must be achieved prior to the start of the project. This phase is described in the VA Tear-Down as the problem definition step of selecting competitor products and preparing in advance of starting the VA Tear-Down project assignment.

FUNCTION ANALYSIS, THE FOUNDATION
OF VALUE ANALYSIS

Function is the end result desired by the customer. It is what the customer pays for. Although function is the requirement, the goal, and the objective, it is not an action; instead, it is the purpose of an action. Function analysis is the cornerstone of Value Analysis. Function analysis is the one discipline that separates VA from the many problem-solving initiatives and processes available to the problem solver. Discovering the power of function analysis, that is, separating the intent or purpose of a thing from its description, then improving value by manipulating functions, is credited to the creative imagination of Lawrence D. Miles.

Some examples of describing functions:

A spring does not move parts; it stores energy.

A screwdriver does not turn screws; it transmits torque.

An oil filter does not clean oil; it traps particles.

Miles used the verb-noun discipline as the rule for expressing functions, and expanded the two-word description by using an active verb and measurable noun to best describe functions. The verb identifies the action to be taken. The noun is the target of that action. In the spring example, "store" is the action; "energy" is the target.

In the VA Tear-Down process, competitor products are disassembled to determine how well the functions are implemented. A function describes what has to be done, not how to do it. As an example, the VA Tear-Down team would not examine the way a spring was designed, but rather how well the design performed the function "store energy" and how effectively that energy was used.

The terms "move parts" for the spring, "turn screws" for the screwdriver, and "clean oil" for the oil filter are the expected outcomes of the functions, and the reason the components were selected to perform those functions. Addressing the functions to be provided, rather than the components selected to perform those functions, broadens the creative boundaries in seeking innovative ways of improving the product. The products "spring," "screwdriver," and "oil filter" are the designs of the manufacturer to provide the functions to achieve the customer's desired outcome.

USING ACTIVE VERBS

When identifying the functions of components, products, or processes, it is important to use active rather than passive verbs. The verb describes the action and the noun defines the object of that action. Searching for the most descriptive verb-noun is difficult. Compromise often results in selecting the action as the noun and using a passive verb to complete the function description.

If a passive description of a function is suspected or you wish to express the function more actively, try to use the noun as a verb and then select another noun.

Wherever possible, avoid using the verbs "provide," "review," or "attend." The verb "provide" is most commonly used by those not practiced in identifying functions. When a function is not understood, the action is used as the noun and the word "provide" is placed as the verb (see Figure 2-3).

Descriptions of functions are taken literally. To use the verb "review" as in the function "review proposals" means read, skim, but do not comment or take further actions. If this is the intent of the function, then the function is correctly stated. However, if the function is more than "review " proposals, the verb-noun statement might be "edit proposals." The use of the verb "attend" will have different meanings dependent on where it is used. For a staff member to "attend meetings" means that the person is just expected to sit there. But for a nurse to "attend patient" means that the nurse is expected to care for the patient.

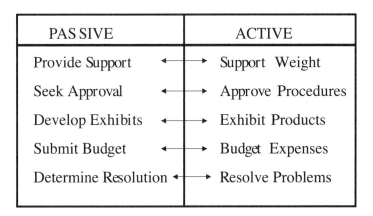

PASSIVE	ACTIVE
Provide Support ← →	Support Weight
Seek Approval ← →	Approve Procedures
Develop Exhibits ← →	Exhibit Products
Submit Budget ← →	Budget Expenses
Determine Resolution ← →	Resolve Problems

Figure 2-3 Passive vs. Active Verbs

MEASURABLE NOUNS

Measurable nouns are easier to determine when the study topic is a hardware product. When hardware components are used, measurements are quantitative and often expressed as engineering units. Examples of measurable nouns include: weight, force, load, heat, light, radiation, current, flow, and energy, to mention a few. Functions such as, "control flow," "reduce weight," and "transmit torque" have nouns that can be universally measured.

In hardware systems, functions such as "repair damage," "complete circuit," and "store parts" have nouns that can be quantitatively measured, but are not easily measurable. "Damage" can be measured in terms of cost or time to repair. "Circuit" can be measured by the size of the network or energy consumed. "Parts" can be measured by quantity or dimensions. Selecting the appropriate measurement for hardware systems depends on the problem to be resolved.

In non-hardware applications such as business processes, functions like "transfer responsibility," "create proposal," and "develop plan" can be measured in terms of time, people, or softer measurement units. Again, the problem to be resolved will determine what measurements to use. Using measurable nouns to describe functions is important for evaluating and selecting the best proposal alternatives that resolve the problem and for presenting proposals for approval and funding authorization.

USING TWO WORDS TO DESCRIBE FUNCTIONS

Miles discovered that the more words it takes a person to describe a function, the less that person knows what that function is. To illustrate this point, when I was in college I was often given essay exam questions. If I didn't know the answer, I could fill many pages in response to the question, hoping that contained in my rambling answer was something the instructor was looking for. However, if I knew the answer, I could express my response in short, succinct sentences.

Miles' genius in defining functions demanded short descriptions, which led to his two-word function rule. Using two-word function descriptions in problem solving is essential because it cuts through technical jargon and creates a communication format that allows members of an interdisciplinary team to communicate with each other. It allows scientists to communicate with financial analysts, engineers with procurement, and manufacturing with marketing. For example, the finance representative on our hypothetical interdisciplinary team, when presenting an idea for consideration, might say:

"Give consideration to obtaining our product at the present time while deferring actual expenditures of capital to a future period."

In time, after some questioning, his suggestion would be understood. Using the verb-noun approach this idea could be expressed as, *"Buy now, pay later."* Although the many subtleties of finance might not be immediately apparent, the team better understands and can agree on what is being suggested.

Miles recognized the difficulty, and sometimes frustration, in trying to find those two words — an active verb and a measurable noun — that could most accurately describe the function of the item under study. The team must then confirm understanding by arriving at consensus of the function description. In his book, *Techniques of Value Analysis and Engineering,* Miles said, "While the naming of functions may appear simple, the exact opposite is true. In fact, naming them articulately is so difficult and requires such precision in thinking that real care must be taken to prevent abandonment of the task before it is accomplished."

DEFINING AND CLASSIFYING FUNCTIONS

The word "function" is commonly used and has many definitions. For our purpose a function is defined as "an intent or purpose that a product or service is expected to perform."

The two operative words in this definition are "intent" and "expected." How products or services are used does not identify their functions. A book may make an excellent door stop, but the function of a book is not to "prevent movement."

In developing the value methodology, Miles classified and defined functions to separate them from their design descriptions. Once defined, functions can be examined and analyzed to determine their contribution to the value equation:

$$\text{Value} = \frac{\text{Function}}{\text{Cost}}$$

The two major function classifications are:

Basic Function: The principle reason(s) for the existence of the product or service, operating in its normally prescribed manner.

Secondary Function: The method(s) selected to carry out the basic function(s) or those functions and features supporting the basic functions.

Many value practitioners prefer a simplified definition that describes a basic function as "anything that makes the product work or sell." Those functions that do not support this definition are classified as "secondary" functions.

Secondary functions are sometimes sub-classified as "required" functions. This describes a function or feature that may not contribute to a basic function but is mandated by a customer as a condition of a sale. The size and layout of a personal computer keyboard, incorporating an emergency brake in automobiles, placing buttons on the sleeves of men's sport and suit jackets, all represent costly features or functions that do not contribute directly to the basic function, but are mandated by the customer. Eliminating or modifying those secondary function items would not affect, and may enhance, the product's performance. However, not satisfying these requirements could result in lost sales.

RULES GOVERNING BASIC FUNCTIONS

There are four rules that govern the selection and behavior of basic functions; these rules are important in selecting functions and classifying them as basic.

1. Once defined, a basic function cannot change.
2. The cost of the basic function is usually a small part of the total product cost.
3. Basic functions cannot sell without supporting (secondary) functions and supporting functions must satisfy the basic function.
4. The loss of the basic function(s) causes the loss of the market value and worth of the product or service.

Rule 1: Once defined, a basic function cannot change.

The importance of this rule is that those functions designated "basic" represent the operative function(s) of the item or product, and must be maintained and protected. Determining the basic function of single components can be relatively simple. As noted previously, the basic function of a spring is to "store energy," a screwdriver to "transmit torque," and a filter to "trap particles." As components join to become assemblies and assemblies combine to create products, determining the basic function becomes more complex.

Is the basic function of a butane lighter to "create heat" or "produce flame?"

The answer depends on the problem or issues to be resolved. If the company that is designing a new butane lighter also produces lighter fuel, and the lighter fluid accounts for a significant part of the business, then "produce flame" would be the obvious choice for the basic function. However, if the designers are a new start-up venture company without invested capital constraints, the basic function would be "create heat," because "create heat" gives the designers more creative freedom in designing their product than "produce flame." For example, if "produce flame" were chosen as the basic function, the cigarette lighter in an automobile would be excluded from consideration.

The same rationale holds in identifying the basic function of a pencil. Is the basic function of a pencil to "make marks" or "deposit graphite?" To "make marks" offers more creative paths to explore than "deposit graphite." You must first determine the problem to be resolved before selecting the basic function. By definition then, functions designated as "basic" will not change, but the way those functions are implemented is open to innovation.

Rule 2: The cost of the basic function is usually a small part of the total product cost.

As important as the basic function is to the success of any product, the cost to perform that function is inversely proportional to its importance. This is not an absolute rule, but rather an observation of the consumer products market. Value Analysis defines value as "the lowest cost to perform that function reliably."

The basic function of a one-dollar disposable lighter, "produce flame," can be reliably achieved with a match costing less than one tenth of a cent. The basic function of a Rolex watch that costs in excess of $20,000 can be expressed as "indicate time." A wristwatch displayed on a self-serve rack can be purchased in a drugstore for less than $25, and perform that function fairly well. People purchase Rolex watches for reasons other than the ability or accuracy in performing the function "indicate time. The value of "indicate time" is $25. Are there other functions and features of a Rolex watch worth $19,975? This leads us to the third rule.

Rule 3: Basic functions cannot sell without supporting (secondary) functions and supporting functions must satisfy the basic function.

Few people purchase consumer products based on performance or the lowest cost of basic functions alone. When purchasing a product, it is assumed that the basic function will perform as specified. The customer's attention is then directed to those visible secondary support functions or product features that determine the worth of the product. From a product design point of view, products that are perceived to have high value first address the basic function's performance, stressing the achievements of all the performance attributes.

Once the basic functions are satisfied, the designers then address those secondary functions that are necessary to attract customers. Secondary functions are incorporated in the product as features to support and enhance the basic function and help sell the product. The elimination of secondary functions that are not customer-sensitive will reduce product cost and increase value without detracting from the worth of the product. Changing customer-sensitive functions will change the customer's value perception of the product, which, depending on how it is done, could have positive or negative effects on sales.

Rule 4: The loss of the basic function(s) causes the loss of the market value and worth of the product or service.

The cost of the basic function does not, by itself, establish the value of the product. Few products are sold on the basis of their basic function alone. If this were so, the market for "no name" brands would be more popular than they are today. Although the cost contribution of the basic function is relatively small, its loss, as illustrated in the lighter and watch examples, will cause the loss of the market value of the product.

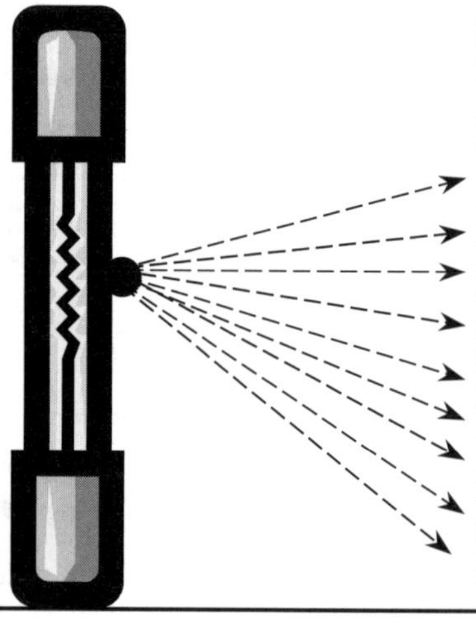

Random Functions	
VERB	NOUN
Break	*Circuit*
Connect	*Circuit*
Protect	*User*
Protect	*Supplier*
Protect	*Equipment*
Identify	*Failures*
Advertises	*Mfgr.*

Figure 2-4 The Functions of a Fuse

FUNCTION EXAMPLE

In Figure 2-4, a number of function descriptions are given for a common fuse. All of the options describe the function of a fuse. The reader's assignment is to select which of these options best describe the basic function. Do any of the functions accurately describe the basic function? If not, can you think of one?

If you selected "break circuit" why not remove the fuse and not replace it in the circuit? How about "connect circuit?" If this is the basic function, why not eliminate the fuse from the circuit and connect the wire ends? Let's consider "protect equipment." If this is the accepted basic function then ideas such as a security fence, guards, and guard dogs could be considered in brainstorming the basic function "protect equipment."

A description of the fuse would be to protect electrical equipment by preventing electrical surges, beyond the circuit tolerance, from entering the circuit and destroying components that are not rated to accept the surge current. "Protect equipment" is an outcome of the fuse's basic function. A better answer would be the function to "limit current." It is that function that was selected as the way to protect equipment, and the fuse selected as the product to implement that function.

The most common response to identifying the function of a home thermostat is to "control temperature." If this were so, we would not need a heating unit or an air conditioner. We could simply mount the thermostat to the wall and set it to the desired temperature. The basic function of a thermostat is to "control current." Its function is similar to a wall light switch. By controlling current, the thermostat turns the heating unit and air conditioner on or off. A supporting function of the thermostat is "sense temperature." It is this function that tells the thermostat when to turn the units on or off.

RANDOM FUNCTION DETERMINATION

The fuse exercise describes the function analysis technique created by Miles. The technique, "Random Function Determination," randomly selects the components of a product, identifies their functions, then determines which of those functions are basic or secondary.

LEVELS OF ABSTRACTION

As seen in the fuse example, a simple product can have many functions among which can be found its basic and secondary functions. If we examine the product in more detail, we can move to a lower level of abstraction and examine its components. The fuse consists of a glass tube, a low electrical tolerance resistance strip, terminal ends, and a bonding media to assemble the product.

Each component of a fuse performs many functions, and on its own level of abstraction, each has at least one basic function. However, many individual basic functions become secondary when moving the focus of the study to a higher level. The glass tube and terminal ends are important to the fuse, but less important when evaluating the contributions of the fuse to the equipment it protects.

The gas-filled cigarette lighter has about 19 parts. Each part, in isolation, has its own basic function. However, these parts and their functions are designed to support the basic function "produce flame." When the fuel is expended and the basic function cannot be performed, the cigarette lighter is thrown away. However, all of the components of the cigarette lighter are still operable. Why then is the lighter thrown away? The answer is in basic function rule 4: the loss of the basic function(s) causes the loss of the market value and worth of the product or service.

COMPONENT - FUNCTION SELECTION

When performing a function analysis on a common lead pencil, the pencil's basic (B) and secondary (S) functions are determined by evaluating the components' contributions to the overall product (see Figure 2-5).

The function "make marks" was selected as the basic function of the pencil; that function is accomplished by the lead in the pencil. Because all the other functions of the pencil are secondary, or in support of the basic function, they are candidates for elimination, consolidation, or modification to reduce the cost of the product. The manipulation of secondary functions is accomplished by the creative efforts of the team, provided that the team does not change the basic function of the pencil. This does not mean that a sintered rod of carbon graphite (or lead) must be used to perform the basic function. It does mean that at the conclusion of the pencil study, the pencil must retain its ability to "make marks" in a manner that satisfies the customer's sense of value.

By selecting "make marks" as the function to address in seeking creative alternatives, the scope of exploration is expanded, because there are many ways to make marks. However, if the client was a lead pencil manufacturer,

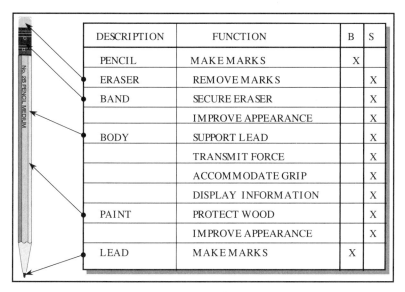

DESCRIPTION	FUNCTION	B	S
PENCIL	MAKE MARKS	X	
ERASER	REMOVE MARKS		X
BAND	SECURE ERASER		X
	IMPROVE APPEARANCE		X
BODY	SUPPORT LEAD		X
	TRANSMIT FORCE		X
	ACCOMMODATE GRIP		X
	DISPLAY INFORMATION		X
PAINT	PROTECT WOOD		X
	IMPROVE APPEARANCE		X
LEAD	MAKE MARKS	X	

Figure 2-5 Pencil Functions

and wished to retain the current way to make marks, expressing the basic function, as "deposit graphite" will narrow the scope of creative investigation.

In Value Analysis, the manipulation of secondary functions is guided by reducing the cost contribution of performing those functions. Eliminating or combining many non-value adding secondary functions and their parts achieves the objective of VA, to improve value by improving the cost-to-function relationship of a product.

In the pencil example, if you make a longitudinal cut in the wooden body and carefully remove the lead, you can write (make marks) with the lead without benefit of any of the secondary or supporting functions. Using just the pencil's lead to make marks would reduce the cost to perform the basic function, but basic functions alone do not satisfy value. To address customer value one must ask; does this represent the best value proposal for the pencil? Only if the proposal is successful in the market place.

FUNCTION COST MATRIX

The Function Cost Matrix approach to performing function analysis is a graphical extension of the Random Function Determination method. The objective of this process is to draw the attention of the analysts away from the

COMPONENTS	COST (IN CENTS)	REMOVE MARKS PERCENT	COST	SECURE ERASER PERCENT	COST	IMPROVE APPEARANCE PERCENT	COST	MAKE MARKS PERCENT	COST	TRANSMIT FORCE PERCENT	COST	ACCOMMODATE GRIP PERCENT	COST	DISPLAY INFORMATION PERCENT	COST	SUPPORT LEAD PERCENT	COST	PROTECT WOOD PERCENT	COST
ERASER	0.43	100	0.43																
METAL BAND	0.25			50	0.13	25	0.06			25	0.06								
LEAD	1.20							70	0.84	30	0.30								
BODY	0.94					10	0.09					40	0.37	5	0.05	5	0.37	40	0.35
PAINT	0.10					50	0.05											50	0.05
TOTAL	2.92	15	0.43	4	0.13	6	0.20	29	0.84	27	0.79	2	0.05	2	0.37	13	0.35	2	0.05

←—56—→

Figure 2-6 Function Cost Matrix

cost of components and focus their attention on the cost contribution of the functions (see Figure 2-6).

The Function Cost Matrix approach displays the components of the product, and the cost of those components, along the left vertical side of the graph. The top horizontal legend contains the functions performed by those components, as determined in the Random Function Determination exercise. Each component is then examined to determine how many functions that component performs, and the cost contributions of those functions. For example, the eraser cost 0.43 parts of a penny, and 100% of that cost is dedicated to the function "remove marks." The metal band, which costs 0.25 parts of a penny, performs three functions: "secure eraser," "improve appearance" of the pencil, and "transmit force" when using the eraser. A cost estimate roughly allocated the cost to perform those functions by estimating the process and material cost of the metal band component.

To determine the cost of the function "improve appearance," the illustration shows that the metal band, body, and paint all contribute to that function. The analyst can now read the chart vertically to determine the cost contribution of performing the function "improve appearance." At this point in the process, it is more important to determine the relative cost impact of the functions than attempt to accurately determine the manufactured cost contribution of those functions. Detailed cost estimates become more important following function analysis, when evaluating value improvement proposals.

Reading across the row marked "total" shows the cost and percent contribution of the functions of the pencil. In this example, approximately 56% of the pencil cost supports two functions "make marks, " and "transmit force." Addressing the cost to perform functions, rather the cost to produce the parts that carry out those functions, will help the VA Tear-Down Team select which areas to brainstorm for value improvement analysis.

LEVELS OF ABSTRACTION

At this point one may surmise that the complexity of a product or system that contains a great many more components than a pencil would be proportionately more complex. Simply put, one could not begin to imagine what a Function Cost Matrix of an automobile would look like after going through the pencil exercise. However, the level of abstraction selected to perform the analysis governs the complexity of the process, not the number of components in a product. Using the automobile as an example, a high level of abstraction could contain the major subsystems as the components under study, such as: the power train, chassis, electrical system, and passenger compartment. The result of the Function Cost Matrix analysis could focus the team's attention on the power train for further analysis. Moving to a lower level of abstraction, the power train could then be divided into its components (engine, transmission, drive shaft, etc.) for a more detailed analysis.

Another approach to simplify the process while maintaining its validity is to select only the basic functions of the individual components for the function part of the matrix. This is a good approach for a simple product that has about 20 to 30 components. Using Random Function Determination, find the applicable functions for each component of the product, classifying those functions as basic and secondary to that component. Then, using only the component's basic functions, display those functions on the Function Cost Matrix. Because the analysis will be moving to a higher level of abstraction by evaluating the product, many of the component's basic functions displayed in the matrix will be secondary to the product. The basic function of a door, to "control access," would be secondary to a house whose basic function is to "create habitat."

To summarize, each component has a basic function. However, that basic function may be considered secondary to the system the component supports, unless the basic function of the component is also the basic function of the system. In our fuse example, the basic function of the element is to "limit current," which is the same basic function of the fuse.

VALUE ANALYSIS OF NEW AND EXISTING PRODUCT

The techniques used in Value Analysis and VA Tear-Down involves a search and evaluation of alternate ways to meet the required performance at the lowest cost. The alternatives available for consideration are greater in a new product (Value Engineering) than for existing products (Value Analysis). Existing products will generally have design constraints; for example, the changed items must be downward compatible with non-changed parts to support after-market services. Existing products will also have cost constraints such as procurement commitments, scrapping or modifying existing inventory, changing service and design documents, modifying tools and fixtures, and editing existing sales literature.

New product development has fewer constraints. It also assures that the cost-benefits occur in the initial production. The break-even point for the investment of a value-engineered new product is far more favorable than the break-even point of products put into service and then modified later for product improvement.

Although the payback for new products is better than for improvement of existing products, the latter must not be abandoned. Existing products have a market that is constantly changing. Cost reduction and value-adding improvements of existing products require constant attention to keep up with aggressive competitors and technology advancements, and to maintain or improve market share and sales. It is in this arena that VA Tear-Down excels.

THE VA TEAR-DOWN PROCESS

The major process events of VA Tear-Down are selection, disassembly, analysis, display, and examination. There are, however, several important steps that need to be addressed before beginning this VA Tear-Down process. The Analysis step in the VA Tear-Down process is the most challenging and is the major factor in achieving the desired outcome. However, one cannot simply jump to analysis without proper preparation.

Most important for a successful tear-down is to follow the process steps. Deviations within those steps may be tolerated to fit the unique needs of a project, but the principles of the building-block sequence of VA Tear-Down should be followed. For example, picking a small sample of one or two parts of competing products will result in biased information. Too large a sample will detract from the discipline of the process sequence and could dilute the desired outcome. The correct number of samples is dependent on the scope of the study, focusing on those target competitors being benchmarked. Getting the maximum value from VA Tear-Down requires completing each step in sequence, in compliance with the objective of each step.

Specific procedure steps with application examples will be explained in detail in subsequent chapters.

BASIC VA TEAR-DOWN PROCEDURE

PLAN

Senior management has a menu of initiatives from which to select a process to capture an opportunity or resolve a problem. The decision to use the VA Tear-Down process depends on achieving a desired outcome by comparing the physical properties of a company's products to those offered by its competitors. A desired outcome may be a lower-cost product that maintains or improves customer appeal.

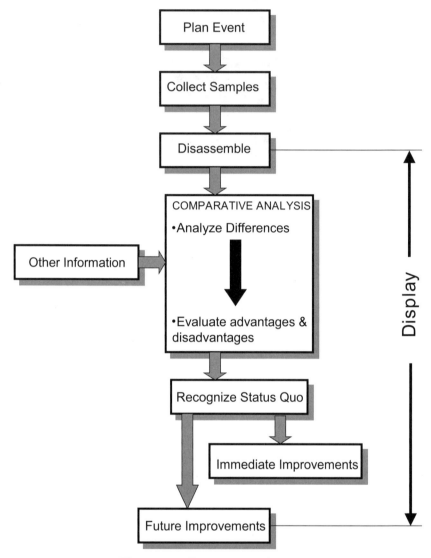

Figure 3-1 Basic steps of tear-down

The choice of competitor products subjected to VA Tear-Down is also influenced by the purposes, organization, and personnel involved in the process. Prior to introducing VA Tear-Down into a company, a management consensus is essential covering the 5W1H questions: why (purposes), who, what, where, when, and how. The "how" question includes how much, or the extent to which the activities are to be pursued. For example, should the analysis cover a full product such as an automobile model, or a subsystem such as a steering system or transmission? In addition, the level of Tear-Down to be employed, the process for analyzing, defining, and implementing improvement suggestions, and the timing of incorporating such improvements all must be established. It is also important that the system for accepting and acting on the suggestions for improvement in a timely manner is well established.

COLLECTING SAMPLES

As mentioned earlier, VA Tear-Down is a comparative analysis process requiring competitor products for the purpose of comparison. Outcomes will vary widely depending on the objective of the project under study and the samples chosen for comparison. It is important to establish study objectives that are within the VA Tear-Down capabilities to achieve.

In evaluating the competitive edge of our own product, it is essential to compare it with market leaders as well as with direct competitors. However, bold innovation requires going beyond these comparisons. To achieve higher levels of technical and process innovation, comparison with products from other industries will be useful.

All these factors must be fully considered in the planning stage in order to collect competing samples and other products that are most suitable for the purpose of comparison.

DISASSEMBLY

Before dissecting the product, the disassemblers should have a clear understanding of what they are looking for. The products collected for comparison are then disassembled into their parts following a "dynamic tear-down" procedure (which will be explained later). In dynamic tear-down, the time required to disassemble each product is recorded. The time it takes for disassembly provides information regarding the buildability of the product. The acts of disassembly and recording the time and sequence occur simultaneously. Unlike other types of tear-down procedures, dynamic tear-down requires that the function of each part of the product be identified and recorded.

Furthermore, unlike conventional tear-down procedures, VA Tear-Down requires that the personnel responsible for the product analysis and improvement proposals comprise the VA Tear-Down team. Much valuable information is lost to the analysts if disassembly is left entirely to a disinterested production team. Skilled production staff members are valuable additions to the team, under supervision and guidance of the responsible VA Tear-Down team analysts.

Information is gained by using all the human senses. Knowing what to look for, the analyst learns much through touch, smell, sound, and sight in the act of disassembly. Simply reviewing the results of another's disassembly efforts loses this source of information. In VA Tear-Down, first-hand experience provides the best information.

If certain parts were designed to perform some function in combination with other parts or assemblies, the analysts may fail to recognize the complementary (or dependency) relation between the parts unless observed first hand during disassembly. As a minimum, members of the VA Tear-Down team, as well as design and production engineers, must be at the site of disassembly to see and record the details of the competitor's techniques.

It is natural to get improvement ideas during the disassembly process. Those ideas should be recorded for consideration during the analysis phase.

The disassembled products are mounted on a display board in exploded view, showing where each part was housed before disassembly.

ANALYSIS

This step in the VA Tear-Down process focuses on analyzing differences, advantages, and disadvantages of the selected product samples.

Analysis is the core of the VA Tear-Down Process. With a general understanding of VA Tear-Down, the team moves to the analysis phase. During analysis, the team observes parts during disassembly, suggests improvements that can be made, identifies differences between our product and competitor's through analysis, and records all such information. Whether to accept such differences as they are, or do something to resolve the differences, is left to later steps. Success of this step depends on the quantity and diversity of ideas raised during analysis.

Even though two parts look identical at a glance, a closer look can identify any small differences. It is these differences that account for one product being better or worse than another. It is, therefore, important in the analysis step to record all the differences, major or minor, in function, structure, materials, and processes.

Equally important to observing and recording differences in the products being disassembled is to discover the reason for those differences. The purpose of VA Tear-Down is not just imitation, but to ask and answer three questions:

"Why are these two products different?" "Are there any advantages to the differences?" "Why was that particular design or process approach selected?" On the basis of the answers, solutions can be found which can make a significant contribution to the company's technology data bank.

When evaluating each difference in terms of advantage or disadvantage, it is not sufficient to just focus on using the results for modifying our products. The timing and sequence for such modifications must be addressed. Issues such as finished goods inventory, investment expenses, immediate competition concerns, and general technology trends must be considered in planning product design changes. Information should be harvested to be used in subsequent Tear-Down steps and shared with those departments responsible for subsequent actions. Specific analysis procedures will be explained later for each phase of the VA Tear-Down procedure.

COLLECTING OTHER INFORMATION

When we put our hand on a competitor's product, we are likely to overly focus our attention on differences that we can observe. In this section, other information does not include the comparative observations that can be obtained from the analysis of a product's functions. Instead, other information includes historical records of product failures, potential structural weaknesses, and the availability of new materials or new processes that may be needed whether or not used on the competitor's product. Ideas about the structure or assembly of products from different industries and how other industries address similar functions would also be helpful. Any related information that helps develop an idea, or information on competitors' strategy, is useful and makes the Tear-Down process more effective.

An analysis of mini-car steering systems made a great contribution to improving the steering of heavy-duty trucks. Applying ideas derived from electric and home appliances to passenger car temperature-control systems also resulted in a cost-effective, functionally responsive, quality solution. Intentionally stepping outside the traditional product boundaries and focusing on the functions of components and systems offers a better creative perspective for developing ideas, which can often translate into a great success.

DISPLAY

In this step in the VA Tear-Down process, the results of our analysis are presented to those most concerned with the product being analyzed. The participants are encouraged to compare the product offering with the competitor's, focusing on the level of technology and competitive advantages and disadvantages of each product. As the participants offer their comments, their ideas and observations are recorded for future consideration.

The display process is divided into steps, or levels. The first step during this process is performing a comparative analysis. As described above, this involves uncovering competitive advantages, then collecting and developing ideas to incorporate changes that offset those competitive advantages. In this stage, the competitive features between products displayed are apparent; thus, the changes that our product must undergo to compete with the competitor's products are identified. A number of options representing different ways of incorporating the improved functions and features are also noted and displayed.

In the second display step, top management are invited to view the display, as well as the analysis and ideas for overcoming competitive advantages. The display room serves as a forum for corporate coordination and high-level decisions regarding strategic positioning of the product under study, developing a sales strategy, preparing a business case justification, and selecting the best ideas for refinement and implementation. If not enough ideas are collected in the first step, or if there is a shortfall of ideas to fit the business strategy, additional idea generation is encouraged and added to the analysis during this second step.

In addition to actual products, data such as photographs, test results, catalogs, customer reactions, and, if appropriate, videotapes are prepared so that visitors can readily recognize differences from observation and reading the published information. This encourages visitors to share their own knowledge about the products being compared. The data must be arranged so that the visitors can visually make comparisons, supplemented by published information rather than having to read the details to identify differences.

Care must be taken not to hide problems in an attempt to protect past decisions. Exposing problems is an essential part of the VA Tear-Down process, but such exposure must be non-judgmentally presented. Otherwise, whatever impact the Tear-Down has on improvement activities are lost.

When inviting people to visit and comment on the displayed products, the invitations should not only include the senior management of the company, but also a broad cross section of interested staff and support members. This includes marketing, engineering, production, suppliers, and focus groups.

Making the display room a forum for sharing information among all company levels helps develop a credible business plan to support proposed product and process changes. Cultivating senior management's appreciation and interest will develop the necessary support for integrating the VA Tear-Down process as a high-level business culture, rather than a mid-level special event.

There are many ways for collecting ideas. Suggestion or idea sheets are available for those who desire to think about and then write out their thoughts. Interviews and recording comments are the best ways to solicit top management's ideas. Technical staff members who have participated in disassembling and analyzing the products displayed are usually those who are directly involved in proposed changes; they tend to be too narrowly focused on the technical aspects. Those visitors who are not directly involved in the proposal development and implementation, or who have a vested interest in the resolution, represent third-party opinions. The views of third-party people represent "outside-the-box" designers, material experts, production engineers, suppliers, and sales people. These people are also valuable sources of information and can contribute unique ideas. All are good supplements to the ideas collected in the analysis step.

RECOGNIZING THE STATUS QUO

Status quo recognition is observing differences, as they currently exist. Without proposing any ideas to close the gap between the products, the observer's attention is drawn to the apparent strengths and weaknesses of the compared products.

In this, the second stage of the display phase, our existing products are compared with competitors' products to determine what weak points our products have and what improvement will close or overtake competition in the current and future markets.

When first observing the displayed products, ideas begin to flow on how cosmetic, product design, and process changes can improve the product offering. The ideas collected, as described above, are then classified by the difficulty and complexity of incorporating the ideas into our product.

IMMEDIATE IMPROVEMENTS

The output from the major steps discussed thus far (disassembly, analysis, display) includes brainstormed suggestions and ideas supported by analysis. Individual ideas that complement each other are bundled into proposals and the timing for implementing the ideas considered. Timing involves questioning whether the ideas should be incorporated immediately into our product, or held for some future date or event.

Two conditions require immediate change and improvement. The first are those value-adding competitive changes that can be implemented with a minimum investment and minor impact on production. The second is the necessity of correcting a serious competitive defect quickly, which could impact production and require a significant investment.

Because the product is in the field and can be evaluated by conducting a market survey among existing customers, market analysis and verification that justify the need for such immediate actions can be performed relatively quickly.

Timing for getting these changes into the market may be a tactical business issue. If so, it may not be prudent to wait for the next scheduled model change when the improvement ideas can be smoothly integrated into the model change process. What can, or must be improved immediately should be done, factoring in a business case justification of potential sales and profit increase offsetting capital expenses and production line disruptions.

Implementing immediate improvement changes carries a risk that must be analyzed against its potential benefits. Once the decision to implement the proposed changes is made, a dedicated implementation task team should be formed to expeditiously address and resolve design and production problems.

FUTURE IMPROVEMENTS

Excluded from "Immediate Improvements" consideration are changes that affect the basic product design. Such changes are highly customer sensitive. They tend to alter the identity of the product under study, which can then be perceived by the market as a new product entry, rather than an improved existing product. The best time and place to consider incorporating "Future Improvements" is when planning for the product's model change.

Model changes or model additions are a recognized and expected market event in the hard goods consumer market. During this event, potential customers compare the attributes and functions of competing products looking for innovative features and compare the prices of such innovations.

In this step, improvement targets are assigned by benchmarking the competitive products selected in the VA Tear-Down process using specific numerical values, or metrics. Improvement targets reflect customer sensitive characteristics such as functions, features, and attributes (e.g., cost, weight, package size, numbers of components, and performance). Using actual competitive data lends credibility to the improvement targets and acts as a creative stimulus in generating ideas that meet or beat the target competition.

Implementation strategy and plans to achieve the improvement targets must consider the number of months or years that the company expects to maintain the change advantages. One cannot assume that competition will be inactive during this period. Change management is dynamic and must consider the competitor's reaction, anticipating when the next round of improvement is warranted.

GENERATING UNIQUE AND MANY IDEAS

The VA Tear-Down differs from conventional teardown techniques in the generation of improvement ideas. Most conventional Tear-Down processes are performed by skilled technicians, in a dedicated "Tear-Down" room, who are not required to participate in the idea generation process. In VA Tear-Down, the staff responsible for resolving the problem issues are directly involved in each step of the process. Their involvement will produce a rich fountain of ideas that are captured, cataloged, and processed.

Figure 3-2 below is a historical assessment of the number of ideas, by percentage, generated during each step of the VA Tear-Down process.

The quality of ideas is more important than the quantity of ideas. However, to get good ideas requires generating lots of ideas.

Ideas are encouraged and collected throughout the entire VA Tear-Down process, beginning when first viewing the competitors' products, prior to the disassembly and display steps. The highest percentage of good ideas harvested and selected for detailed analysis occurs during disassembly and analysis. During these steps the responsible designers, manufacturing specialists, and cost analysts are directly involved in the disassembling and analyzing of the products. They are the most qualified and motivated to assess and analyze the differences in the displayed products, and to create ideas and determine their value-added characteristics.

However, as noted above, ideas are also encouraged from the casual observers and the management staff invited to "walk through" the display area. Although the yield of good ideas from casual observers is relatively small

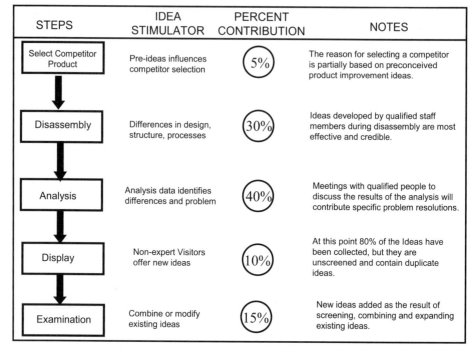

Figure 3-2 Numbers of Ideas Generated During Each Step of the VA Tear-Down Process

(see Fig. 3-2), those accepted make a positive contribution to improving the product. Putting a limit on ideas, or excluding those people not directly involved, is likely to miss good ideas that may be conceived from an entirely different perspective.

To avoid duplicating ideas, a short description of the ideas offered is posted on the display board to be viewed by the casual participants. Such lists also stimulate new, or improved ideas by "piggy backing" on these ideas.

JUDGMENT AND SELECTION

Finally, we come to the step where we examine, judge, and select the good ideas that have been collected thus far. Selecting may suggest just accepting or rejecting those ideas. What is essential to this step is to select acceptable ideas, and justify rejecting ideas.

It is human to reject 95 percent of the goodness of an idea for the 5 percent that is wrong with it. When an error or fault is found in an idea, no matter how small, it tends to taint the entire idea. To overcome this condition, judging the idea should begin with the statement, "Yes, if ..." instead of the normal "No, because...." This statement guides the evaluator to use constructive criticism in judging the ideas.

As an example, instead of saying, "No, the idea is no good because management will not approve the required investment," try saying, "Yes, the idea is good, if we can justify the investment to management." Using this positive approach to judgment and selection makes the "required investment" an issue to be resolved, rather than the rationale for rejecting the entire idea.

As noted in Figure 3-2, justifying idea rejection also leads to new or a combination of ideas.

SELECTING PRODUCTS FOR COMPARISON

There are a number of factors to consider in selecting products for comparison in the VA Tear-Down process. Prior to selecting products for comparative analysis, the reason for the VA Tear-Down project should be re-examined and the desired outcome confirmed.

The primary objective of VA Tear-Down is to uncover the reason for a competitor's advantages. Therefore, selecting that competitor's product and other similar products for comparison is obvious. Because the target competitor is known, simply looking at the product before disassembly will generate ideas. Figure 3-2 indicates that 5 percent of the improvement ideas generated result in this preliminary comparison stage. However, if only the target competitor's products are selected, the best that can be hoped for is a continuous catch-up process.

To overtake and maintain the lead over a competitor's product requires thinking out of the box. Translating the design and process differences into function descriptions, and including products for comparison that provide the functions, can achieve this objective. As an example, if a fog lamp for an automobile were selected for improvement, the comparable lamps used by our competitors would be analyzed in a series of simple, established Tear-Down steps. Taking it to a new level, the functions needed in fog lamps, as distinguished from other lamps, would be identified. Using the function descriptions as a guide, other lamps from a wide range of industries would be selected for comparison in addition to competitor fog lamps.

The state of technology should also be considered in selecting products for comparison. There is little value in selecting older products that have not been upgraded for a number of years, even if they are still enjoying successful sales and profits. A product using thick and thin film technology cannot be compared with older products still using semiconductors or axial lead electronic components. Today's watches have solar cells and washing machines have built-in microcomputers. Rapid changes in material, design, and process technology are offering unique methods for meeting affordable and profitable function improvements.

The primary products selected for comparison and analysis are those competing with our own in the same model class and market. In order to collect as many ideas as possible, the process shown in Figure 3-3 is recommended.

Innovative improvements require ideas, and ideas are hard to develop unless there are some creative stimuli. In VA Tear-Down, the initial stimulus comes from the competing products chosen for comparison. Going beyond the norm is achieved by choosing a wide variety of additional products, not just those similar to those that we want to improve.

SELECTING PRODUCTS WITH SIMILAR CONFIGURATIONS

As stated previously, products that directly compete must always be included in selecting products to be compared and analyzed. This is basic to the Tear-Down process. If we are to wrest market share away from our target competitor, we must learn all we can about its product, process, and market strategy. In many sections of this book, the concept of comparison with directly competing products is stressed. A case can be made for cultivating ideas only derived from comparison with direct competitors. This approach leads to ideas for overcoming current competitive disadvantage and this alone will lead to a partial success if catch-up is the objective.

To stimulate innovative ideas requires looking beyond the limits of examining the target competitor's product. Looking at product offerings from different producers in the market sector as well as products from different industries will be needed. However, searching for additional products should be guided by selecting products with functions and characteristics similar to those of the products being evaluated. Venturing outside of direct comparison with the target competitor will expand the list of improvement ideas. Ideas are cultivated in the process of sorting out differences in those common characteristics

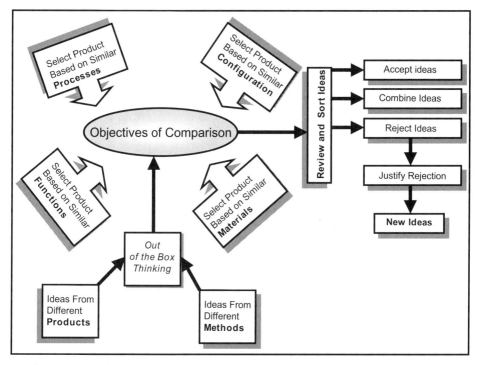

Figure 3-3 Guide For Selecting Competing Products For Comparison

SELECTING PRODUCTS WITH SIMILAR PROCESSES

A selected manufacturing process can be used in producing a variety of products for diverse applications and industries. In improving the design of a fog lamp, for example, there is much to be learned from harness connections in plastic parts forming, assembly method, etc. Some plastic parts are finished in two-tone color; others have a thin color film bonded to its surface. In a casting, the shape, finish, material, and production quantity of the final product design will influences which casting process to select.

Even though similar processes make similar products, changing the processes by modifying the design of the product often results in a significant cost reduction. In the highly competitive consumer hard goods market, design engineers are keenly sensitive to the economic impact of their design decisions on the cost of the product they design. That sensitivity is based on being familiar with a wide variety of manufacturing processes. This process information, applied in analyzing the processes used to produce the selected samples, can uncover and counter competitor's process advantages.

SELECTING PRODUCTS WITH SIMILAR MATERIALS

This selection option calls for collecting for comparison products from a wide range of products, based on their materials, regardless of their use or functions. The objective is to find any advantages in the use of materials. If our product improvement candidate is made from plastics material, examining different plastics products will provide information on several ways to use that material. Exploring how the material is reinforced (with glass fiber, carbon fiber, etc.), how it is colored (hot stamping, two-tone color, etc.), and how it is formed for small volume production (vacuum forming, blowing, etc.) will stimulate improvement ideas. Selection of samples will be more effective if it is recognized that different kinds of products that employ similar materials contain ideas that are not present among products with the same kind of use or function.

SELECTING PRODUCTS WITH SIMILAR FUNCTIONS

Searching for product examples by function is a very effective way to break through traditional boundaries. Although products may share the same function, their configuration, process, materials, application, and other design characteristics could differ widely. Therefore, comparing the fog lamp to Christmas tree lamps, street lamps, and household lamps could be useful in exploring new improvement ideas. All of the examples share a common, basic function: Emit Light. The purpose of the function, to illuminate an area, is common to all the examples. Products also emit light in different ways. Some color light, amplify light, focus light, modify the light spectrum, or manipulate light in other ways.

Selected candidate lamps could also include flashlights, neon signs, highway lights, and others. Comparison focus could include lens design, color spectrum, reflectors, bulb design, assembly techniques, wiring schemes, or the use of materials, among others. The way other industries achieve similar functions in their use of materials, design, and processes could be quite different from those considered normal practices in the automotive industry. That is why isolating selected functions to look outside paradigm limits can provide valuable insights beyond the competitor's product for improving the fog lamp, or other products.

COMPARISON BY METRICS

This method of comparison differs from those described above and can be used in association with other comparison methods. Comparison by metrics, such as cost per kilogram, cost per component, or cost per unit weight, can reveal the source of product cost problems that result from the design, tooling, or process. It is also interesting to use this method to gain a better perspective on the metrics commonly used within an industry. For example, using cost per unit of weight, some may be surprised to find the cost per kilogram of a washing machine is higher then the cost per kilogram of a Mercedes automobile.

MANAGING IDEAS

The differences that are found as a result of comparison and analysis are the source of ideas. When gathering ideas, judgment should be suspended. Judging and evaluating ideas is a separate and distinct step that begins after a reasonable quantity and variety of ideas have been sorted and recorded.

Resistance to ideas is a natural and human trait. When presenting ideas for comment, such resistance, if properly managed, can be a positive opportunity to modify, qualify, and reinforce those ideas.

Design engineers invest considerable time on research and development before their design is finalized. It therefore takes unusually objective engineers to view ideas originating from less-than-credible sources in a dispassionate, analytical manner. It is difficult enough for them to keep up with technology advances in their own field. Therefore, when ideas are presented for improvements that do not originate with their peers, the natural reaction is defensive and skeptical. However, ideas by marketing, customer sources, manufacturing specialists, and representatives from different industries are sensitive to faults and design shortcomings as well as improvement opportunities. Their ideas are not meant as criticism, but as offers of assistance in meeting the goals of the Tear-Down assignment.

An event separate from generating ideas should be established for judging ideas. The meeting format must accept rebuttals and criticism, but criticism must be constructive. In this setting it is okay to question the value and practicality of ideas. But the objective should be to find and convert ideas into a proposal and a plan, rather than rejecting ideas without just cause. A successful Tear-Down can result only from using ideas extracted from differences found from a wide range of viewpoints and various types of analysis and evaluation.

ESTABLISHING VA TEAR-DOWN TARGETS

The primary reason for establishing targets is to set business objectives to which the staff responsible for their achievement can relate. Targets also define success and should be expressed in quantifiable terms. A project target described as, "Create a best-performing product at the lowest cost," leaves the terms "best" and "lowest" subject to wide interpretation. Thus, it is ineffective as an improvement goal.

SETTING COST TARGETS

Setting cost targets is a common and important metric. Product cost targets are selected based on business needs, strategy, and objectives, and are influenced by market conditions.

Although price is influenced by cost, the two respond to different business conditions. "Cost" is the expenditure of resources required to manufacture a product that satisfies customers' sense of value. "Price" is the value manufacturers places on products that anticipate how much customers are willing to pay for the functions and features offered by that product. A successful business is one that can satisfy customer requirements at a cost that results in planned profits.

Setting price is a strategic decision left to the executive management of a company. Cost is the consequence of a design, the manufacturing process, and the creative talents of those who are responsible for the design and manufacture of the product.

In a seller's market, where the demand for the product exceeds its availability, more attention is given to the product's profit opportunity than cost of producing the product. Under these conditions, competitive pressures have less influence on products' cost targets.

Figure 3.4a "Seller's Market" describes how the sales price is established in a seller's market condition.

In today's global market, driven by accelerating technology and the ever-growing capacity to produce products, companies compete for sales and market share. Competition is the primary factor that creates a buyer's market, establishing market-driven cost targets. Accelerating technology also creates demand for new products, which reduces the market life of products and increases the need for rapid new product development.

As shown in Figure 3-4b the market, driven by competitive pressures, sets

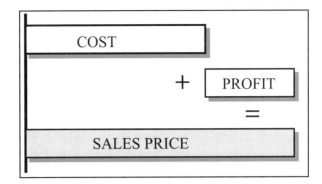

Figures 3-4a
Seller's Market
(Profit Driven)

Figures 3-4b
Buyer's Market
(Market Driven)

Figure 3-4c
Cost Must Be Reduced
to Obtain Profit

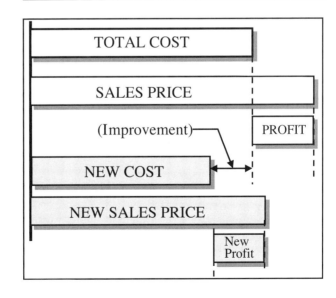

the product price. The company gages its competitive position by establishing prices for similar products in relation to market prices.

Should the sales price be too high for market expectations, a reduced product cost target is established to reduce sales price while protecting profit margin. This is a condition illustrated in Figure 3-4c

Another business condition describing the need for continuing product cost improvement is shown in Figure 3-5. A mature product that has been available for a few years tends to increase in cost as labor and material costs increase beyond the benefits from learning and other continuous improvement activities. Also, newer products force price reductions of mature products to maintain their sales level and forestall product obsolesence. Maintaining the viability and extending the life of the mature product depends on the ability of the company to price compete while returning a profit level that justifies the product's continued existence.

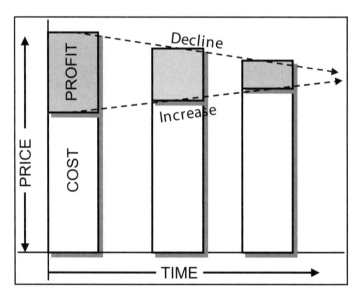

Figure 3-5   Costs Rise and Price Lowers Over Time

In many cases, as described in Figure 3-6, new products must be priced lower than their predecessors, even with improved functions and upgraded features. This condition prohibits adding the cost of the improved or added functions to the current market price. A new target price is required, allowing the incorporation of new and improved functions and features at a lower or

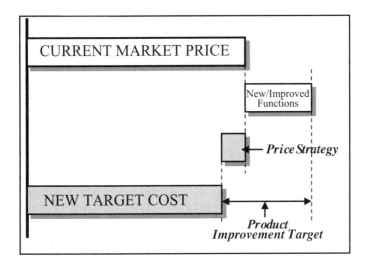

Figure 3-6   New Product Priced Lower than it's Predecessor

equal price than the product predecessor. A case in point involves personal and laptop computers. Technology is developing so rapidly that price per feature is declining at a rate that a computer is almost obsolete on the day of its purchase.

Cost targets for new products must be aggressive. Such targets must be based on the company's ability to keep up with changes in technology, and the advancements made by existing and newly emerging competitors. Start-up competitors are in a better position to take advantage of new manufacturing technology because they are constrained by neither investments in older capital equipment nor the need to extend product obsolesence to pay for product development.

Figure 3-7 Cost and Target Gap

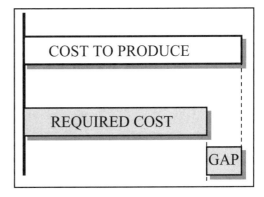

Figure 3-7 illustrates the gap that exists between the cost targets that can reasonably be achieved and an aggressive cost target needed to maintain or improve position in a highly competitive global market.

IMPROVEMENT ACTIVITIES

Product improvement is a dynamic, continuing, cyclical venture. Even with improvement under way as an activity, the advantages of improved products erode as new competing products are introduced into the market. As a result, sales prices must be lowered to offset competitive pressures, which in turn causes profits to shrink, forcing renewed product improvement activities.

Figure 3-8 summarizes the process and product improvement activities employed for setting targets in VA Tear-Down.

In today's hard goods consumer market, it is difficult for manufacturers to increase the price of improved existing products even though material and labor costs have risen. An exception is in the consumer electronics. New product introductions can enjoy a higher introductory price for the producer who beats competitors to the market and can appeal to that market segment that

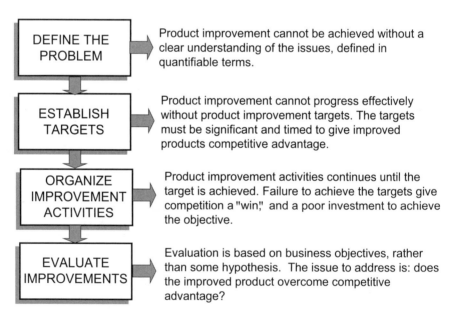

Figure 3-8   Flow of Improvement Activities

wants to be the first to own the "latest and greatest." However, this price advantage is short lived as competing products enter the market.

As manufacturing and engineering technology accelerate, the time from product concept to introduction rapidly decreases. It was not too long ago that manufacturers could look forward to a five-year product life before competition reduced the price and eroded market share. In today's market, by the time a product reaches the market, the need to reduce cost and develop a replacement product begins.

At the 2002 Society for Japanese Value Engineering Conference, held in Tokyo, Japan, the Director, Communications Systems Group, Sharp Corporation, said that the company needed to produce a new product every eight months to keep ahead of competition. The products must not only be attractive and affordable to the customers targeted, but the product must be introduced quickly to beat competition with product introductions. The first product in captures the major market share. That is Sharp's business strategy. The speaker was describing how Sharp developed a digital camera integrated into their cell phone using Value Analysis techniques. The user can take a picture then send it directly to someone through email, send it to another cell phone as a picture, or send the picture home and deposit it in a computer file folder.

Sharp is also very much involved in extending the life of existing products by reducing costs to protect sales and maintain margins.

The VA Tear-Down may not bring back a five-year product life, but it can provide a quick relief method to improve the short-term price competitiveness of mature products by incorporating product improvements that exceed current competition.

SETTING COST TARGETS

Cost targets may include an overall cost target for the product (product cost and investment), and element cost targets for each part, tools, tooling, etc. In VA Tear-Down, it is a common practice to try first to achieve targets for each part of the product, then to sum up individual achievements. Shortfalls or over targets are normal and the negatives and positives are balanced as long as the overall target is achieved. In other words, it is the top target that is important, not achieving the components that total the top target. Three types or levels of cost targets will be discussed: simple, reasonable, and aggressive.

APPLYING VA TEAR-DOWN TO SETTING TARGETS

The basics of VA Tear-Down are disassembly and analysis. Therefore, our products and those of our competitors are first compared in terms of cost. The cost difference (cost delta) is used as the basis for setting cost targets. Cost is the most commonly used target, but anything that can be competitively compared, such as weight, performance, or number of parts, can be used as targets. Performance is determined by measuring control items (e.g. output, capacity), by testing, or by analyzing published specifications. In case of weight, each of the parts making up the product under analysis is individually weighed and targeted.

A tabletop oven toaster is used to explain how targets are established. There are three types of oven toasters made by three different Japanese home electronic appliances manufacturers shown in Figure 3-9. This example was chosen because every Japanese household is likely to have at least one similar toaster and almost everyone can relate to the example. The tabletop toaster oven is also relatively simple in structure and principle; it is easy to understand. Each product can be switched from 400W to 800W and each can be used for a very wide variety of cooking. The retail price of these products is approximately ¥5,000 (Approximately $33.00).

In Table 3-1, the column titled "OURS" is the base product being addressed for improvement. Columns A and B are competitive products. For this target-setting example, Table 3-1 compares the weights of these toasters as broken

Figure 3-9
Picture, Three Brands
of Oven Toasters

No.	COMPONENTS	A	OURS	B	Minimum Wt. In Grams – (Difference from OURS)	Individual Targets
1	Cord Assembly	102	112	80	80(32)	△ 32=80
2	Heater Circuitry	252	287	**203**	203(84)	△87=200
3	Grill Assembly	92	79	**76**	76(3)	△ 9=70
4	Door Assembly	**325**	338	357	325(13)	△38=300
5	Reflector Assembly	415	454	**375**	379(79)	△84=370
6	Switch Panel Assembly	44	110	**30**	30(80)	△ 70=40
7	Cover Assembly	787	762	**716**	716(46)	△62=700
8	Tray	316	**202**	274	202(--)	△22=180
9	Other Assembly	10	**0**	12	0(--)	-- = 0
	TOTAL WEIGHT	2,343	2,344	**2,133**	2,017(337)	△404=1940

Note: 453.6 grams = 1 pound   [1]   [2]   [3]   [4]

Table 3-1   Toaster Ovens

down by their common parts. This evaluation form allows the comparative analysis of the three products. Targets for improving our product are established on the basis of these data. This is an example of target setting in our VA Tear-Down procedures.

SETTING SIMPLE TARGETS

The targets that are established as a result of Tear-Down analysis are those based on physical properties rather than targets based on projections, theory, or market desires. Using physical comparison makes setting targets more credible and persuasive, a more motivating challenge for people who are engaged in product improvement.

Looking at the results of the comparative analysis in Table 3-1, product B is the lightest in total weight at 2133 g ([1] in the Table). This weight becomes the first target and is called a simple target because it can be achieved by simply copying B Co.'s design. It is, therefore, relatively easy to achieve consensus and have everyone agree that "what B Co. can do, we can do." This is the start of ours challenge.

When comparing the weight of each part, note that some of our parts weigh less than product B's parts. If we use those lighter parts, we can achieve an additional 103 g weight saving, which would result in the new product weight of 2030 g. This is our second target and represents the minimum weight target. This would still be considered a simple target because it can be achieved simply by making the modifications as shown in the example.

Let us assume that as product B is being analyzed, aggressive improvement activities are currently under way at B Co. What if they introduce a new toaster that is lighter in weight and cheaper in price than their current product? Considering that the assumption may be valid, we should set targets strategically.

SETTING REASONABLE TARGETS

If we benchmarked the products by combining the lightest parts of the three products, the resultant total weight would be 2017 g ([2] in Table 3-1). Assuming a relationship between a part's weight and its cost, the lightest and probably the least expensive parts should be our target. The weight saving to be achieved on our toaster is 337 g [3] or 14.4 percent. This value is the theoretical minimum; it is our reasonable target.

Setting targets to this level is considered relatively easy to achieve because, with some compatibility modifications, incorporating the lowest weight existing parts will achieve this reasonable target. This level of achievement is attainable, but not considered very challenging.

SETTING STRETCH TARGETS

Establishing challenging improvement targets occurs in setting stretch targets. This level combines the theoretical minimum with potential to establish aggressive and challenging targets.

Competitive products selected for VA Tear-Down analysis and target setting are usually mature products. As noted above, it must be assumed that a competitive replacement product is currently being developed. To move ahead of the competition requires information about advances in materials and processes, market growth, and the progress of in-house technology to assess our product advantages and limitations. These represent the potential, which is factored into the theoretical minimum. In other words, stretch targets involve factoring in unproven potential improvements for each product component, rather than using the sum of the parts in today's market. We set component targets because material and process technology may differ among product parts. The targets established this way set our stretch or strategic target for the product.

This is a credible target because it has been established on the basis of actual product and part examples as well as justifiable potentials and market projections. In the toaster example, the total of the individual parts targets, which reflect the development team's challenge, added up to 1940g [4]— a weight saving of 404g or 17 percent. The product target is 1940g, or the sum of the targeted parts, realizing that in the process, individual part targets may exceed or come in under their assigned targets. It is important to aggressively pursue the product target of 1940g.

Regardless of the target being pursued, whether unit cost, development time, or some highly valued market attribute, the assignment of the target must be based on credible analysis, not established arbitrarily. Targets that are not well founded are likely to turn out to be targets just for the sake of setting targets. In contrast, those that have been established using a rational method such as VA Tear-Down are likely to be more accepted as worthy and challenging targets.

The next step is to determine if the targets established by this process will be acceptable in the market when the new or improved product is introduced. If the market evaluation indicates the need for further efforts, each part in each component will be reviewed to decide first on which challenge targets to focus and second on how much additional improvement is required. The product target is modified based on those component targets singled out for special effort, rather than simply adjusting the product target based solely on the market shortfall. Unlike conventional target setting, which is established on a rough estimation basis, this method shows the parts to be given priority and special attention. This concept is called "feasible cost planning."

Setting targets solely on need makes target setting less persuasive than targets set by VA Tear-Down.

THE VA TEAR-DOWN ORGANIZATION

As technology advances, job assignments become more specialized and located in various parts of the organization. Introducing a product to a market requires the participation of many organizational elements working in harmo-

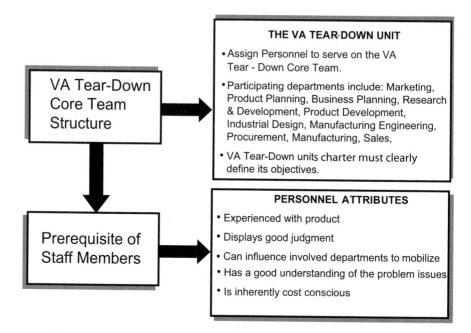

Figure 3-10 Characteristics of the VA Tear-Down organization

ny. These organizations include: marketing, product planning, business planning, research and development (R&D), product development, industrial
design, manufacturing engineering, procurement, manufacturing, sales, and
many support services. In addition, we are in the era of globalization where
similar products are produced and marketed throughout the world.
Individuals or small units don't make meaningful improvements on their own.
Their achievements are limited to the boundaries of their unit. Product
improvement and innovation require collaborating effectively, systematically,
and efficiently across many contributing organizational units.

Individuals can indeed make use of the Tear-Down concept and VA technology to some extent, but a well-established organization is needed if we want to:

1. Apply the VA Tear-Down to a wide range of products or systems

2. Make meaningful successes

3. Deploy VA Tear-Down activities systematically

4. Use VA Tear-Down for corporate coordination

5. Use VA Tear-Down for training and education

6. Apply the technology in other fields

We have thus far described the VA Tear-Down process and how it can contribute to a company's business objectives. The questions now emerging are: Where in the company do you place VA Tear-Down? What organization and at what level in a company is best suited to take responsibility for this process and manage it as a successful venture? Plans for establishing a VA Tear-Down unit in a company should include the qualifications needed in the assigned staff and the placement of the unit in the organization. When considering the staff, consider that multidisciplinary teamwork is essential to produce meaningful results. Figure 3-10 summarizes the criteria for staff selection in building a VA Tear-Down unit.

SELECTING THE PARENTAL ORGANIZATION

Ideally, the VA Tear-Down unit should be corporate-wide and have the ability to reach across other organizations for the talents and disciplines needed for the VA Tear-Down process. There are many candidates for housing this unit. But consider that once the unit belongs to a certain department, the VA Tear-Down activities are likely to be influenced by the charter and influence of that department.

Some candidate organizations for the VA Tear-Down unit include:

1. Planning: Assigning VA Tear-Down to the Planning department has the advantage of it being in a high management-level position that is involved with many company functions; VA Tear-Down could then be deployed easily. On the other hand, Planning tends to be too far removed from the working level where VA Tear-Down methods will be used.

2. Product Design and Development: If product design were a dominant part of the company's business, this would be a viable candidate to house and support the VA Tear-Down unit. Ready access to engineers is a plus in this business structure where product improvements can be easily and reliably developed and implemented. One disadvantage in a company with a high ratio of purchased parts is that it is difficult to involve parts suppliers without support from the purchasing department. Involving the manufacturing department is also important, assuming that the company produces the components of the products it designs.

3. Materials: VA Tear-Down can be placed here if the outsourcing ratio is high, that is, if the company relies heavily on its supplier base to produce the components for its products. Changes for making improvements may take time unless the unit can establish a close relationship with other departments in the company, especially Product Design and Development.

4. Manufacturing: Placing the VA Tear-Down unit in the Manufacturing department should be considered for a company with a high "make" ratio. This placement would ensure timely feedback from field service. That information could be analyzed and used to produce value-added product characteristics. Again, a close link-up with Product Design and Development and parts suppliers is necessary to ensure timely design changes and market responsiveness.

5. Business Management: This is a preferred location for the VA Tear-Down Unit since Business Management is the center for all corporate activities.

THE STRUCTURE OF THE VA TEAR-DOWN UNIT FOR LARGE COMPANIES

A large company is defined as a corporation having multiple strategic business units (SBUs) and divisions within the SBUs producing a range of products.

How to structure the VA Tear-Down unit is dependent on many considerations, including the unit's department affiliation, the size of the corporation, the company culture, management level sponsorship, and available resources.

A STAFF-BASED VA TEAR-DOWN UNIT

People are selected from departments concerned and are assigned exclusively to the VA Tear-Down Unit on a full-time basis. As a closed unit, it will perform all of the management, administrative, and technical functions of a department, from planning to sorting out improvement suggestions. The unit

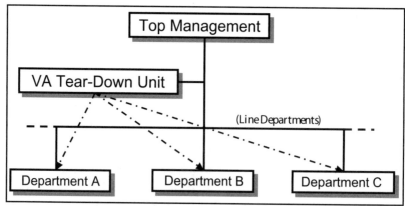

Figure 3-11 Top Management Based VA Tear-Down Unit

may report directly to top management or to some suitable senior manager (see Fig 3-11). The unit coordinates and manages whatever needs to be done by line departments to implement the outcomes of the VA Tear-Down unit.

Some advantages of the staff-based organization structure are:

1. The unit has ready access to top management.

2. Improvement activities can be carried out on a company-wide basis.

3. The assigned staff members can focus their full attention and activities on the assignment rather than be concerned with the duration of the assignment and the work left undone at their home base.

4. As experience is accumulated, the personnel involved become trained specialists, which in turn further improves the effectiveness of VA Tear-Down.

However, a top management organization is not free from some disadvantages such as:

1. The departments from which people are drawn to serve in the VA Tear-Down unit may lose needed skills.

2. Coordination with line operations may be difficult.

3. It may be difficult to deploy activities across the company because of limited availability of the VA Tear-Down team.

A PROJECT-BASED VA TEAR-DOWN UNIT

In this scheme, a small core team is placed in a selected department as shown in Fig. 3-11a. When a product improvement opportunity is identified and sanctioned, a project task téam, consisting of members picked from those departments concerned, is formed to address the project issues. The team addresses the problems on a project assignment basis and is disbanded as soon as the issues are successfully resolved. Each member of the team, chosen for their particular specialty, belongs to the department from which he or she has been selected and contributes the time necessary to complete the assignment. Overall the small core team does coordination.

Some advantages of a project-based organization are:

1. A large number of people can participate in the activities and gain experience with VA Tear-Down.

2. Team activities can easily be transferred to line operation.

3. Multiple project teams can address different projects simultaneously.

Conversely, some disadvantages to a project-based unit are:

1. Team members are not exclusively assigned to the task. As a result, their activities as a team may be limited due to the workload pressures in their home departments.

2. Assigned team members may be inexperienced in the VA Tear-Down process, making the project objectives more difficult to achieve.

3. This structure works well for small, short-duration projects, but it is not suitable for large-scale, major projects.

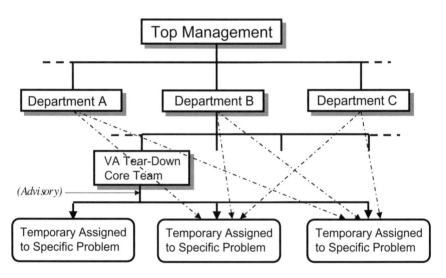

Figure 3-11a   Project-Based VA Tear-Down unit

THE VA TEAR-DOWN UNIT FOR SMALL AND MEDIUM SIZE COMPANIES

The VA Tear-Down method is based on competing product comparison and analysis. As a product improvement process, its application is not limited only to big companies. The concept can and often does achieve profit-improving results in small companies too. However, small and medium size companies tend to be short of human resources, which makes it difficult for them to assign full-time personnel to build a fully-staffed VA Tear-Down unit. The VA

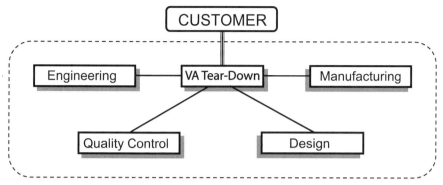

*Figure 3-11b   VA Tear-Down organization in small and
medium size companies*

Tear-Down unit in a small or medium size company will, therefore, be placed at company headquarters, instead of elsewhere in the organization. This location will give the unit the needed contact with markets and suppliers (see Figure 3-11b).

In small and medium size companies, business management convenes a VA Tear-Down task team whenever there is a need for new products, product improvement, or other strategic product change. Team members are drawn from concerned department personnel, who return to their jobs following the completion of the assignment. Additionally, customers and suppliers who have a vested interest in the assignment are invited to participate in the VA Tear-Down task team.

Involving customers and suppliers produces effective results in a relatively short time. Communications with market and supplier sources are more effective, being direct and participatory in the VA Tear-Down process. Feedback is shortened and more credible, reducing the time necessary for investment approval and implementation of the team's recommendations.

CONSIDERATIONS FOR ESTABLISHING A VA TEAR-DOWN UNIT

Each of the VA Tear-Down organizations discussed above has its advantages and disadvantages. When considering a VA Tear-Down unit, the unit must complement and match the company's size, markets served, available personnel, and business culture.

It must be stressed that support and participation by top management as well as incorporating the function of the unit into the company's business process are essential for assuring a meaningful and successful VA Tear-Down unit. Also essential is the investment of suitable resources, including personnel, training, time, facilities, and a budget.

Setting-up a VA Tear-Down Room

VA Tear-Down requires a base for efficient operation, an area and permanent location called the VA Tear-Down room. The facility using VA Tear-Down technology will assure benefits for the company, serving as a center for training personnel and establishing marketing strategies. Important to a Tear-Down room are its location, facilities, size, equipment, and personnel. The room should be open to employees interested in improvement activities, which in turn will help boost employee commitment and morale.

Location

The VA Tear-Down room is the base where products are analyzed against competition, and marketing strategies formed. All of the VA Tear-Down procedures—disassembly, analysis, display, evaluation, and follow-up—should be carried out in this dedicated room.

The room should provide easy access to those who are most concerned with products and market strategy. Usually the people who visit the facility most frequently and stay there the longest are those from the design and engineering departments. Visits from customers and suppliers should also be considered when selecting a location. A separate, stand-alone facility is best to encourage people to visit the facility.

Depending on the characteristics of a company, a VA Tear-Down room can be designed and used as a product show room or located in the purchasing department so that it can be used for discussion and negotiation with suppliers and customers. Where access is limited for reasons of confidentiality, security precautions may be warranted.

Facilities

A Tear-Down facility must provide an environment conducive to study and analysis. Attention should be given to lighting, sound suppression, parts and equipment storage, etc., where people can take time to view and evaluate the displays. Meeting rooms for discussions, analysis, or planning activities should be included in designing the facility. Providing a library containing computers for data access, competitive information, price information, cost tables, and catalogues is a valuable asset for the VA Tear-Down room. Means for outside communication such as telephones, fax systems, and Internet link-up through personal computers are also necessary.

Lavatory facilities and a small kitchen area for breaks and lunch meetings might be considered to enhance the VA Tear-Down Room.

To sum up, essential VA Tear-Down room features include:

1. Reasonably arranged floor, walls, ceiling, and illumination to achieve a bright, positive environment

2. A place isolated or insulated from manufacturing and other distracting noise

3. Equipment and tools for use in disassembly, analysis, and display

4. An area with audio-visual and computer equipment where meetings and research can be conducted

5. A break area where people can relax while viewing the various product displays

6. An area where related VA Tear-Down functions and applications can be taught, including Value Engineering, Design for Manufacturing and Assembly (DFMA), Target Costing, and other initiatives

Size and equipment

The size of the VA Tear-Down facility varies greatly with industry as well as with objectives. An automaker that displays several complete vehicles needs a huge space, compared with a manufacturer of hand-held electronic communication equipment.

A Japanese automobile manufacturer has one of the largest Tear-Down facilities, where car models can be driven in and out of the facility. The Tear-Down facility displays 14 vehicles, each in three different states of assembly; whole vehicles, broken down into blocks (to show the stages of parts assembly), and broken down into every part. The room contains 42 vehicle sets as well as many assemblies such as new axles and transmissions and new model bodies. The displays include the company's products as well as products from their most active competitors. The facility is used for VA Tear-Down activities and for new product development and test.

Fig. 3-12 and Fig. 3-12a shows examples of automakers' and automotive parts manufacturers' Tear-Down facilities, and Fig. 3-13 is a diagram showing an ideal automaker Tear-Down lab arrangement. The size of the room depends on the sizes (from light trucks to heavy-duty trucks) and the number of the vehicles to be displayed, and how many vehicles are to be displayed simultaneously.

Figure 3-12 & 3-12a
Examples of Automobile VA
Tear-Down Room

Figure 3-13 shows an example of a well-designed VA Tear-Down room for the automotive industry. This does not mean that all such facilities should be large structures. VA Tear-Down rooms should be sized to fit the needs of the company, its marketing strategy, and the products they produce. A strategy to consider is to start modestly, allowing the successes of the VA Tear-Down activities to justify a larger facility. This approach appeals to conservative managers who believe investments in facilities should be justified by their resultant output.

VA TEARDOWN ROOM TOOLS AND EQUIPMENT

A VA Tear-Down facility needs the proper tools and equipment. VA Tear-Down is more than a disassembly and inspection process. The objective of VA Tear-Down is to analyze competitor advantages in the market place, not to

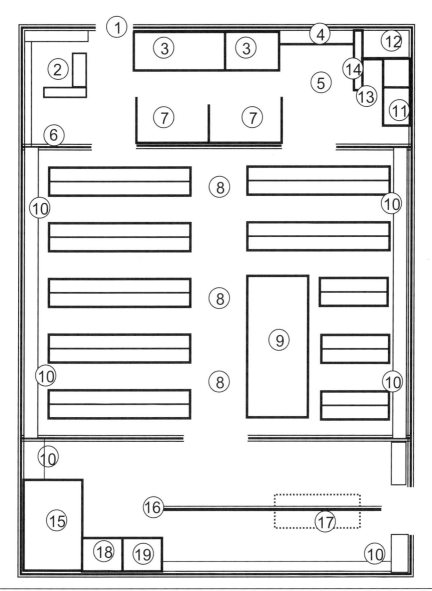

1. Entrance 2. Reception 3. Meeting Room 4.Material Space 5.Free Space 6.Information Display 7.Team Design 8.Main Display 9.Large Components 10.Parts Shelves 11.Measuring Instruments 12.Lavatory 13.Kitchenette 14.Copy/Print Equipment 15.Tool Room 16.Hoist 17.Floor Pit 18.Wash Stand 19.Power Supply

Figure 3-13   Automobile VA Tear-Down Room Layout Example

emulate those functions and attributes, but to leapfrog over competitors. This requires a level of equipment and tooling commensurate to the depth and levels of required analysis and to be able to display the results.

In a dynamic VA Tear-Down, events are often videotaped while time is being measured with a stopwatch. The video is helpful in subsequent analysis and verification. Such devices are affordable and well developed for recording (optical and digital cameras, video recorders) and for reproduction (TVs, personal computers) and should be included on the tool and equipment list.

Displaying the results of the VA Tear-Down process is necessary to support the requested investment for product improvement.

The following discussion looks at typical tools available in a VA Tear-Down room. It is presented as an example and should be modified to fit each company's situation.

Disassembly Tools

Disassembling tools vary from industry to industry. Screwdrivers and box wrenches will be sufficient for products assembled mostly with bolts and nuts. Welded products will require grinders, files, and chisels. A universal tool kit should also be included.

For Dynamic Tear-Down, which simulates line operation, air tools and electric tools should be available. During the start-up of a new VA Tear-Down facility, renting equipment to reduce capital expenses is an option.

Labels

When more than one product is disassembled, it is difficult to identify the source and name of the components. Different color labels for each product can avoid confusion. Each label should be a printed form to enter information such as part description, weight, or cost.

Measuring Instruments

Sizes (dimensions, sheet gauges), weight, paint thickness, and surface roughness represent most of the attributes measured. Some products may require large scales or balances. For more exacting measurements, calipers, micrometers, push-pull gauges, a gloss tester, and perhaps a microscope may be required. Digital instruments are becoming popular. A good selection of instruments will make the VA Tear-Down operations more effective.

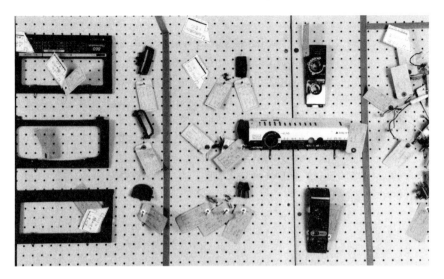

Figure 3-14   Perforated board

Display Board
Boards are used for displaying what has been disassembled. Parts are hung and data are posted on the board. A perforated board (see Figure 3-14) or mesh board is convenient for this purpose.

Tables
Tables covered with colored paper protected by a transparent vinyl sheet help identify the display, making the display more inviting to observers.

Posting Displayed Items
Displays should be designed so that visitors can hold and observe the item being displayed. Displayed items should be hung or placed on tables rather than tied down.

Tapes, String
Colored tape, wool yarn, or string can be used for dividing the space and for connecting parts and assemblies displayed with their explanations or descriptions. The purpose of the display is to help visitors quickly and correctly understand what is being displayed.

Writing instruments

Marking pens and other writing instruments should be available to create posters labels and signs. A wide range of sizes and colors will create an attractive display. Color-printed information produced on personal computers is also a good option.

Suggestion sheets

Visitors are requested to turn in proposals or suggestions. Prepared suggestion sheets, focusing on the objective of the tear-down project and strategically placed at the display, will encourage visitors to present their comments. The completed forms should be collected in a binder identifying the display.

Recording equipment

The use of tape recorders available at the display will help collect suggestions from visitors who have an aversion to filling out forms.

IDEA CONTROL AND FOLLOW-UP

In Value Analysis (VA) and Value Engineering (VE) workshops, a large number of ideas emerge during the stage of idea development. Many ideas are offered in brainstorming, cutting across a broad range of project topics. These ideas grow through a maturing process into specific proposals by filtering through several stages of sorting, discussion, and investigation.

The brainstorming process used in VA/VE to generate ideas can be made more effective through the VA Tear-Down process. One feature of VA Tear-Down is that it is directed to specific objectives for specific products. Ideas are raised on the product's functions and attributes that relate to the expected product improvement outcome. This procedure narrows the field of investigation, but improves the quality of initial ideas and the development of more effective proposals.

CHARACTERISTICS OF IDEAS COLLECTED THROUGH VA TEAR-DOWN

In VA Tear-Down, ideas tend to be more practical, if less creative than those generated through brainstorming because the ideas result from examining actual products and associated hardware. This practice often confines the team to the limits of existing technology. Although focusing on current products tends to stifle thinking out-of-the-box, it does have pragmatic advantages.

Because the emerging ideas are based on actual examples, they are easier to evaluate, understand, and explain.

Because proposed models can be constructed in the VA Tear-Down room, credibility and confidence in the proposal is increased through appeals to all the human senses.

Even if an idea proposed is one that tests and breaks present boundaries, departing significantly from the current product, the proposed idea has the advantages of an actual, physical model.

TARGETING AREAS FOR IDEA DEVELOPMENT

The idea development team addresses the quality and quantity of ideas collected as part of the VA Tear-Down process. This team discusses how each process step influences where to direct idea gathering and how to collect and combine those ideas.

COMPETITOR SELECTION

In selecting competitors for the VA Tear-Down comparative analysis, the question "Why was that competitor's product selected?" should be addressed.

This competitor has cost advantage. The basis for the cost advantage can be discovered and will be the source for generating ideas for cost reduction.

The competitor's product claims advantages in the functions offered. If these functions have market appeal, they will be a focus of idea generation.

A competitor offers a new or improved product. A product analysis will determine what is known about the product and what innovations have been incorporated. Those innovations will be the subject of idea development.

THE SPECIFICS OF DISASSEMBLY

During disassembly all differences, whether advantages or disadvantages, are noted. Everything in which the competitor differs from our product is the source of ideas, regardless of advantage or disadvantage. The competitor's advantage suggests ideas for product improvement; its disadvantage exposes competitor product weaknesses. Each difference helps to evaluate competitor potential.

Everything that is not found in our product, such as the way of assembling, tool fixtures, parts mating, quality assurance, or the way parts are fastened, is also the source of ideas.

Particular attention should be placed on tooling and tool fixtures as well as the parts themselves. Production tooling is usually standardized at each company, and is not subject to frequent changes. Because tooling designs differ between competitive companies, analyzing the type of tools and fixtures used offers a good opportunity to identify advances in manufacturing technology.

Product Development should be part of the disassembly team. Many ideas can emerge during the disassembly and dissecting of competitor products and they should be added to the growing list of ideas.

In the course of analyzing a competitor's new product, some new information that has not been noticed in earlier models is often discovered. Record everything whenever such an item is noticed.

HOW TO COLLECT IDEAS

In the process of collecting ideas, take photographs or draw sketches when the product is partially disassembled. Add the appropriate explanation to the photographs or sketches.

Whenever a specific suggestion or proposal is offered, record it on a suggestion form. Also note the existing design from which the idea developed, in addition to the suggestion or proposal.

When disassembling causes parts to be deformed or destroyed, making identification difficult, the product must be preserved in its partially-assembled state for subsequent review.

Quantitative as well as qualitative analysis should be performed during disassembly.

The disassembly team should include experts from specialty fields. Plastics engineers, for example, may suggest replacing pressed parts with plastics parts whereas casting engineers may suggest thin-wall or precision casting to eliminate parts and reduce assembly. These experts raise the largest number of ideas during disassembly.

Supplier representatives who have experienced some problems in the past may find a similar problem in the competitor's current design.

Standard suggestion forms must be readily available so that visitors and other participants can record their ideas at the time they are noted.

A suggestion form or idea memo must have the name of the suggestor as well as the product and area proposed for change. This is necessary for tracing the suggestion to its source whenever any question or additional information is needed.

To minimize the number of casual ideas or duplications, it is advisable to have ideas that have already been proposed in this and previous steps posted with the displayed product. Another advantage to posting ideas is to piggy-back, or expand on the ideas presented.

SCREENING IDEAS

When screening ideas, recognition should be given to the suggestor of the idea. The person who turned in an idea thinks his or her idea will contribute to product improvement goals. That motivation should be maintained and encouraged

If an idea seems unacceptable, identify why or what problem would result if the idea were implemented. Consider modifying or incorporating the idea with another to make it acceptable.

Changes over time, such as changes in legislation, regulations, environmental concerns, materials, customer requirements, and future trends, must always be considered when screening ideas. These factors may change design and evaluation standards, which the competitor may have incorporated in its product.

Acceptable ideas that cannot be implemented now should be investigated to determine what additional actions are required for implementation, or when the idea will become acceptable. Such ideas constitute a valuable data bank for the future (model change or any change in environment).

BEYOND IDEA COLLECTION

Product improvements can only result from implementing acceptable ideas. Simply suggesting ideas only increases the knowledge of competitor advantages. The ultimate objective of VA Tear-Down is product improvements that will beat competitors' offerings. This objective can only be achieved through the full use of those valuable ideas.

1. Make the examination of the collected ideas visually friendly. Ideas and the affected parts should be organized and arranged on display boards and tables.

2. Sort the ideas by the departments from which the ideas have been proposed.

3. Table displays should contain a title card indicating a specific area of the product being analyzed.

4. Provide additional suggestion forms where examination results can be entered

5. Schedule target dates for implementing approved suggestions.

6. Information should be displayed showing expected benefits including cost, quality, function, features, and attributes as well as estimated investments.

INITIATING IMPLEMENTATION

Once the analysis and idea selection for product improvement have been completed, the implementation plans that show areas of responsibility, major milestones, and critical paths are developed. A control table for each part must be created, then shared with those people concerned with the implementation phase. The table should contain the following information:

1. Accepted idea

2. Implementation timing (final dates for completing drawings, testing, engineering release, and implementation in production)

3. The designated Project Manager

4. Effects of implementing the idea, including those on investment and payback

5. Implementation and progress reports

The success of the VA Tear-Down process is dependent on the implementation of those accepted ideas. Product improvement implementation should be treated as a fully endorsed and committed business venture led by a project manager. Progress reports should be scheduled as part of the manager's staff meeting agenda.

Without effective implementation there can be no product improvement benefits.

CHAPTER 4

APPLYING VA TEAR-DOWN TO ISSUES OF CONCERN

The previous chapters described the history, theory, and process of VA Tear-Down. This chapter will focus on issues of concern, or themes.

Conventional Tear-Down methods rely on observation and analysis to uncover those differences in selected products that account for a competitor's advantage. VA Tear-Down relies on dissecting the product to discover the changes in design that contribute to the value-added characteristics of the competitor's product. The result is a proliferation of suggestions covering a broad range of issues that must be sorted, classified, and analyzed. The VA Tear-Down Process is used for this competitor analysis approach, but determining the issue of concern (or theme), setting objectives prior to beginning the VA Tear-Down process, and focusing the VA Tear-Down process on those themes will make Tear-Down much more effective.

At the completion of activities addressing themes, a comprehensive determination of the competitor's potential for product improvement, at present and in the future, will emerge.

In this chapter, VA Tear-Down procedures for each theme are explained in detail and are arranged in outline form so that readers can make practical and easier use of the information. Further reference is made to procedures, analysis methods, tools, and document forms associated with the selected Tear-Down process.

To begin using VA Tear-Down, readers should follow the procedures described in this chapter as closely as possible. As experience accumulates, the process and methods can be modified to satisfy the company's particular needs, markets, processes, products, and culture. It is important to stress that VA Tear-Down and its application addressing issues of concern, are not limited to the products and examples shown in this book. The competitive position of any product or process can be improved using VA Tear-Down, if that product or process can be dissected, functions and features determined, and differences displayed for analysis.

DYNAMIC TEAR-DOWN

Most products consist of many parts that are assembled in various detailed, sub- and final-assembly operations. Dynamic Tear-Down applies the principle of comparative analysis to the process of assembly. The term "dynamic" refers to all the design features that contribute to the time and cost of assembling the product in production. Some of these features include hand selecting parts to avoid tolerance build-up, the need for adjustment during assembly, the number of individual parts needed to perform a function, and the need for special assembly tools.

In Dynamic Tear-Down, following disassembly, the products are re-assembled as intended by the product's producer. The sequence of the assembly is noted and the time required is recorded, simulating the producer's assembly process. The sequence and assembly time are compared, and the function of each part is determined to assess its contribution to the product.

SCOPE OF APPLICATION

Dynamic Tear-Down is applicable to any product consisting of components that can be assembled and disassembled.

OBJECTIVES OF DYNAMIC TEAR-DOWN

The objective of Dynamic Tear-Down is to reduce the labor and capital expense required to assemble a product. The assembly operation is a major labor cost component and a prime target for product improvement by reducing fixed cost. The following are some Dynamic Tear-Down characteristics.

Reduce Assembly Time

Comparing the ability to assemble products with that of competitors' products reveals advantages and disadvantages for our assembly process. The object is to overcome competitor assembly advantages by incorporating similar or better reduced-time assembly procedures. Implementation is effected in two steps: improvement of current products, and when either the next model change occurs or a new future model is developed.

Determining Competitive Advantage

Because the ability to assemble is determined by measuring time and by comparing and analyzing the products under study, the competitive advantages of our product and the competitor's products can be determined quantitatively.

Improve Fit and Finish Quality

Comparing and analyzing how competitors secure their assemblies can establish advantages in reliability and performance. This is determined by analyzing the means to prevent fasteners from loosening after assembly. The focus of this issue is the tolerance of parts, locking devices, and method of positioning the parts for assembly.

ESSENTIALS OF DYNAMIC TEAR-DOWN

Lacking specialized tools, equipment, or trained personnel used in assembly, and not having the detailed assembly procedures, it is difficult to replicate the way a competitor assembles their products. Competitor products are, therefore, disassembled using the tools, personnel and techniques employed in our facility. The team should:

1. Disassemble the portion of a competitor's product that is similar to its counterpart in our product, and assemble that portion several times until the assembler is familiar with the assembling process.

2. Once a degree of assembly proficiency is achieved, measure the time required to assemble the components.

3. Similarly, assemble our product and compare the time with their assembly line, and determine the difference between the assembling time of the competitors' product just measured with the time on our assembly line, in percentage. That percentage can be used for projecting the competitor's assembly time on the assembly line.

Basic Steps

The basic steps of the Dynamic Tear-Down process are shown in Figure 4-1. The temptation to eliminate some steps or take short cuts should be avoided, particularly if the VA Tear-Down team is unfamiliar with the process. Modifying the process to reduce time could result in a less-effective outcome. This is true of any type of Tear-Down. Procedures in each step are explained below.

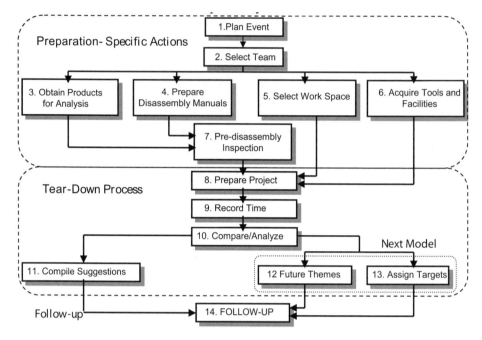

Figure 4-1 Basic Steps of Dynamic Tear-Down

PREPARATION - SPECIFIC ACTIONS

Preparations for carrying out VA Tear-Down follow a set of planning steps called "5W1H." 5W1H refers to the sequence of asking, and responding to five questions: Why, What, When, Who, and Where. The 1H refers to How. The answers to the questions affect the entire VA Tear-Down process and should be followed as a planning outline for a variety of market and product applications.

In the 5W1H planning steps, "What" refers to our products as well as the competitors'. Answering "What" determines the objective of the Dynamic Tear-Down project and the disassembly procedure. A completely different Tear-Down theme and process output will result depending on the answer to the "what" question. "Who" and "when" are equally important. Results will vary with the disciplines selected for the Tear-Down team. Specialists tend to seek and analyze areas involving their specialty. Timing for implementing the recommended actions must be effective. Speed and output are very dependent on the proficiency and experience of the Tear-Down team conducting the analysis. It therefore follows that 5W1H must be carefully planned for maximum success.

1. Plan Event

Why: Why is this action being taken? The objectives of the Dynamic Tear-Down project must be clear as to why that project was selected and the expected outcome of the project. The Tear-Down project team cannot achieve the project's objectives unless there is team buy-in and consensus regarding the project's objectives and outcome.

What: Decide what product to work on. The reasons for selecting products for Tear-Down analysis include: direct competitor, looking for technology superior to ours, desire to change the market price, and perception of worth. Also consider when the competitors' products will be available for the Tear-Down project.

When: Develop a schedule. The VA Tear-Down plan should include an events schedule with process milestones. The schedule should allow for sufficient preparation and lead-time. Other schedule considerations when planning the VA Tear-Down project include; tooling preparation, prototype work, tests, drawing release, implementation timing etc.

Good planning practices are essential to achieve the desired results of the project investment. Process performance, as scheduled, adds credibility to the process. Conversely, schedule slips and unplanned events affect the team and management's confidence in the VA Tear-Down process.

Who: Who does the work? This planning step includes the selection of team members, and identifying those responsible for producing the improved product. Also consider assigning staff specialists and, in the case of a purchased product, technical representatives from the supplier.

Depending on the scale and goals of the project, it may be necessary to recruit team members from many departments concerned with the outcome of the VA Tear-Down project.

The Dynamic Tear-Down involves a number of related assignments. The roles of each team member and his or her contribution to the project must be defined.

Where: The site for conducting the Dynamic Tear-Down could include the parent company, the principle supplier's facility producing the parts, or an off-site location most convenient for the gathering of the selected team members.

How: How to carry out a Dynamic Tear-Down? This should be considered in two ways: What Tear-Down method best complements the issues of concern (in this example, Dynamic Tear-

Down), and what target is established for product improvements?

2. Select Team

The team members chosen in this step are the working level personnel in the Tear-Down process. Organizational considerations are discussed in Chapter 3 (see "The VA Tear-Down Organization"). Many of the assignments involved in Dynamic Tear-Down require the services of specialists. The outcome of Dynamic Tear-Down is largely dependent on member selection. Core members required for Dynamic Tear-Down include the following three groups.

A. Assembly personnel: Engineers who can reproduce each step of assembly on the production line.

B. Time-recording personnel: Time for every motion or step in the line operation is measured and recorded. This job requires the services of a qualified industrial or manufacturing engineer.

C. Recording personnel: Each disassembly step is recorded. This involves the use of computers, video, photography, and other audio-visual equipment. Specialists in software graphics and photo imaging should participate and record the details of disassembly and assembly of the products being analyzed.

When selecting specialists to serve on the Tear-Down team, it is best to arrange the schedule to accommodate a qualified staff member than opt for a less qualified person.

3. Obtain Products for Analysis

Plan carefully and follow-up to ensure that the competitor's products are delivered and are available for VA Tear-Down when required. In the automotive industry, competitor parts and products received are put through a variety of tests and data collected after they have been obtained. The same attention should be given when drawing parts from our production line. Arrange for receiving the parts well in advance to avoid conflicts with scheduled customer orders.

4. Prepare Disassembly Manuals

Disassembly manuals are not necessary for simple products. Complex products such as automotive subsystems require some information regarding their parts, structure, and arrangement. Competitor maintenance manuals and parts lists help the Tear-Down team understand the products, and operation procedures help determine the sequence of the Tear-Down disassembly operations.

5. Select Work Space

In the event that a permanent VA Tear-Down facility is not available, it is necessary to allocate space temporarily, for the duration of the project. Ideally, the disassembly Tear-Down is conducted on or adjacent to the assembly line on which our product is produced. However, the actual assembly line or adjacent space is usually not available because of interference with the production line and differences in the competitor's assembly conditions. The best compromise is to arrange the allocated space to simulate the assembly operations as closely as possible.

Some considerations for selecting a temporary area for Tear-Down include:

>   A. Enough space for the disassembly and assembly operations, recording, and necessary investigation.
>   B. Good illumination for the operations.
>   C. Air pressure and power for the disassembly tools.
>   D. Hoists and lifters for heavy products (if required).
>   E. Temporary depot for tools and parts storage.
>   F. Meeting space for team members to analyze and discuss the product once it is available.

6. Acquire Tools and Facilities

An example of a tool list usually used for the Dynamic Tear-Down is shown at the end of this section. For specific applications, decide which tools to prepare for simulating assembly by consulting the team members. Major tools and facilities to be prepared include:

>   A. Tools and equipment required for disassembly and assembly, including benches.
>
>   B. Recording instruments such as stopwatches, process analysis tables, cameras, and videos.
>
>   C. Storage area and cabinets for labels or tags, parts shelves, dollies, supports for heavy components, etc.
>
>   D. Equipment for meetings such as tables, chairs, and white boards.
>
>   E. Other equipment including personal computers for data input or design-for-assembly (DFA) analysis on site. DFA and its relationship to VA Tear-Down is explained later in this chapter.

7. Pre-Disassembly Inspection

With the Dynamic Tear-Down team in place and the products (ours and the competitor's) and related information available, the Tear-Down process can begin. The objective of pre-operations inspection is to plan specific steps of the Tear-Down with the products physically in front of the team. This is also an opportunity to inspect and investigate the products in assembled form. The team is familiar with their product, but unfamiliar with the competitor's products. This is the time for gathering information that would be lost once the disassembly begins.

During this pre-operations inspection we should look through the eyes of the customer in examining the products to determine value perceptions, or why the products have customer appeal. By examining the products in their assembled state, we examine and compare the products through the eyes of the customer to uncover value perceptions. Value perceptions include overall look, size, weight, ease of operations, and packaging.

TEAR-DOWN PROCESS

As in the medical profession preparing for an invasive procedure, make sure that all the tools and equipment needed are laid out to follow, as closely as possible, the planned disassembly sequence. Make any needed adjustments in the process or schedule. Photograph the competitor's product before disassembly, focusing on value perception areas. During disassembly, record any sequence changes with video or photographs.

Next, collect the necessary performance data. Even if catalogues and specifications that show quantitative information are available, it is important that actual values (output, performance, etc.) are confirmed by measurement and test. If all the products are not available as scheduled for Tear-Down, measurements can be taken simultaneously with the disassembly operation. However, preliminary investigation and verification is best done prior to disassembly.

1. Prepare Project

The Tear-Down team is usually familiar with the assembly operation of their own products, but competitors' products are not necessarily similar in structure or assembly steps. Furthermore, the team leader of the Tear-Down project is not necessarily the same person responsible for assembling that product on the line. Therefore, the Tear-Down team should become familiar with their assigned products prior to beginning the formal disassembly procedure.

The following steps describe how best to become familiar with the products selected for analysis:

A. Repeat disassembly and assembly several times beforehand.

B. Decide on the order of assembly steps. Assembly can be accomplished by using the additive method or the subtractive method. In the former, parts are assembled from the ground up, as on a production line, fitting one part to another until the complete product is assembled. In the subtractive method, only those parts that we need for time measurement are removed and then replaced after measuring the time for the operation. When time measurement is complete, remove just those parts, which have been time measured, and then remove and replace other parts on the product.

C. Select the proper and best tools for our operation.

D. Prepare a good environment (e.g., bench height, illumination, power source).

E. Prepare assembly order list by displaying how parts form detailed assemblies, how the detailed assemblies fit into subassemblies, etc., for each model of product. This sequence will make time measurement in subsequent process steps easier.

F. The degree of completion of a product that is subjected to Tear-Down should match that in which such a product is used as a part on our own assembly line. If purchased parts are part of the product, then the part as it has been received from its supplier is disassembled.

G. Repeat disassembly and assembly several times to minimize differences in time resulting from variation in familiarity.

Assembly must simulate, wherever practical, the order in which our product is assembled on their production line, because the order of assembly on competitor's line is not usually known, or obvious.

2.   Record Time: The Core of the Tear-Down Operation

Time is best measured after we have developed a level of proficiency and familiarization with the product, its parts, and the disassembly process. We are ready when assembly and disassembly can be performed smoothly and confidently. Time for certain steps begins when a part is taken from where it is stored. Time for a person to bring that part to where it is needed and time for assembling operation are measured. Measurement ends when the assembled part is properly placed and ready for the following steps.

A. Recording must be made correctly and precisely. Items to be recorded are: time, motions, parts, tools, weights of the parts used in that step, types and number of "fixing" parts, and the ease or difficulty of operations in that stage. Fixing parts, which are a category of parts used in assembly, include bolts, screws, clips, rivets, and other fastening devices.

B. By means of photographs, video, or sketches, record advantages, disadvantages, or unique competitor characteristics of the assembly method.

C. When taking a photograph or videotape, make sure the part or product being recorded is labeled and has a scale showing its relative size to avoid any later identification errors.

D. Time, in the standard form of labor-hours for their product. should be determined so that the same units of measure can be used in subsequent evaluations.

E. Tear-Down is a comparative analysis method. In cases where assembly operations can reasonably be deemed identical, our product labor-hours can be applied if they are comparable to the competitor's product. Comparative analysis focuses on differences, not comparable operations and time.

F. Time measurement must be sorted and compiled within the day the measurements are taken, while the memory of those who took the measurement is still fresh. Recalling the details of operation becomes more difficult with time. If any issues arise, try to resolve them by the end of the day.

G. After completion of the day's work, discuss the outcome with the team and review the collected information.

3.   *Compare/Analyze*

This step begins the analysis of advantages and disadvantages found in disassembling the products. The level and precision of the analysis has a direct effect on identifying competitive advantages and the development of ideas to overcome them. Particular attention should be given to the method of mating parts, their finish and fit, and achieving their level of quality:

A. Create a Labor-Hour Analysis Comparison Sheet.
Compare and record those factors that account for differences in assembly and disassembly of parts. Using photographs, sketches, and parts lists as reference, describe and illustrate differences in labor-hours, types and numbers of fixing parts, difference in process, and ease or difficulty in operation.

B. Create an overall report.

"A picture is worth a thousand words." The comparative analysis report should be visually dominant, focusing on displaying the differences in products by highlighting advantages and disadvantages. Use figures, graphs, and photographs whenever possible; minimize words.

Essential to the overall report are labor-hour comparisons, factor analysis, comparison of the number of parts used, comparison of the types and numbers of fixing parts, detailed comparison of problems involved, overall evaluation, and other competitor assembly practices. All these observations need to be analyzed and recorded.

C. Process caution-oversight in transferring data.

To ensure complete and accurate reporting it is best to collect data during disassembly, rather than at a later time, when relying on memory. Any lapse in data collection or errors in interpolating the information are lessened when all the products are available for analysis. However, if a critical entry or a team analysis is missing, data errors will occur resulting in a faulty or incomplete report.

Sketches in the labor-hour comparative analysis form, in addition to a list of parts, will enhance the data by illustrating why differences in labor-hours resulted. This will support the conclusions and recommendations in the report.

4. Compile Suggestions

Any item for which comparison has revealed differences is a possible source of improvement suggestions. Clues for improvement are taken from the labor-hour factor analysis in the Labor-hour Analysis Comparison Sheet, or from the overall report's summary.

These differences are translated into potential improvement suggestions and priorities assigned on the basis of the benefits and feasibility of each suggestion and estimated time to implement. The suggestions are reviewed and refined into a Dynamic Tear-Down project team proposal, backed by an implementation action plan. The proposals are then submitted to the project authority for implementation action.

5. Future Themes

Summarizing what has been learned to this point, the Tear-Down team has completed an in-depth inspection of competitors' products; it is familiar with the competitor's technological trend as well as with how they fit, mate, and fasten parts. The team has recorded product assembly time and determined the competitive advantage in the analysis process. The current theme or issue of concern includes, for example, our product's ability to pass the competitor's inspection standards and the differences in state-of-the-art technology. The results may lead to the evaluation of other issues of concern.

6. Assign Targets

Each Tear-Down theme should have clear objectives in terms of quantitative targets for the issue being addressed. The targets will be useful in plotting progress during follow-up. Improvement targets should be set higher than the competitor's performance, for two reasons. First, it is not enough to catch up. The objective of VA Tear-Down is to pass competition. Second, when evaluating the competitor's product, it must be assumed that they are also actively seeking product improvements beyond their current models.

FOLLOW-UP

In this step, checks are made at regular intervals to make sure that the approved proposals are being implemented on schedule and on target. Submit correction action reports to senior management soliciting any needed support. If progress is falling behind, find the cause for delay and take corrective actions. There can be no product improvement benefits unless the proposed and approved actions are implemented.

To maintain management's interest and priority, it is suggested that the expected outcome be reported in management terms. The outcome of the Tear-Down project may result in lower cost, a lighter unit, or a product easier to operate and maintain. But how do these product improvements affect company goals such as increased margin, internal rate of return, potential sales, return on assets, and other business success indicators?

EXAMPLES OF ACTUAL APPLICATION

Figure 4-2 is a simple, but good example of Dynamic Tear-Down. In the bottom row in the picture are a common bolt, washer, and nut. The top row illustrates a "Sems Bolt," that is, a ready-made combination of a bolt and a spring washer. Using Sems eliminates the time required to install the washer. The Sems Bolt in the upper right corner has a tapered tip, which makes the assembly or attaching operation much easier than the flat-point bolts shown in the bottom left corner. Shown at top center is a flanged nut, which, like the lock washer below, prevents turning, or backing out and loosening though vibration. Although the fasteners look similar, the Sems Bolt is easier and faster to assemble by hand, and a better design approach for automated or Design-for-

Figure 4-2
Bolt and Nut Display

Assembly (DFA) consideration.

Several types of bolts and nuts are displayed in Figure 4-2. Those in the top row reflect efforts to reduce assembling time and improve reliability.

Figure 4-3 shows an example of the Labor-Hour Analysis Comparison form. A similar form compared how the feet are installed on the bottom of each toaster. The feet and their fasteners are shown in the photographs of Figure 4-4. These small parts reflect their designers' ideas about their length and about how to keep the toasters from skidding when their switches are activated.

Some Essential Dynamic Tear-Down Points

Below is a summary of key Dynamic Tear-Down points:

1. Determine the total number of system parts, as well as the number of attaching parts (fasteners). Time required for assembling is about proportional to the total number of parts.

2. Find advantages in the competitor's design that contribute to good assembly practices. Such inherent design features assure good quality.

3. Record the procedures with video and other visual devices to make subsequent review and research easier.

Tools Used in Dynamic Tear-Down

Although Dynamic Tear-Down tools were mentioned previously, they are listed again for reinforcement.

1. Tools for collecting information: Catalogs, parts lists, operating instructions, maintenance manuals

2. Tools for assembly: Electric and air tools, benches, illumination, hoists (lifters, chain blocks, jacks for heavy products)

3. Tools for recording: Stopwatches, cameras, videos, drawing boards, measurement instruments, personal computers, etc.

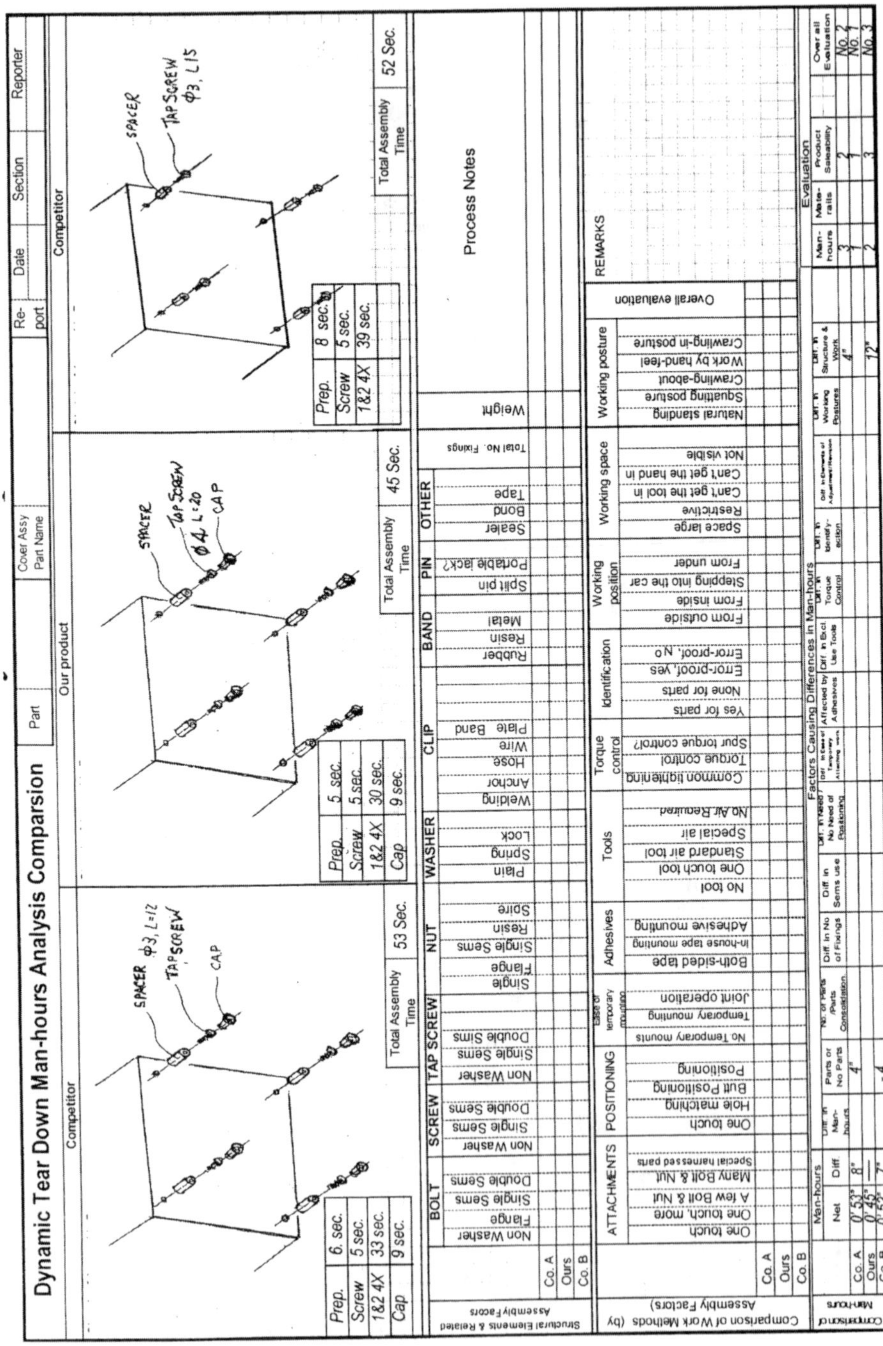

Figure 4-3   A Labor-Hour Analysis Comparison Form

4. Tools for storing: Labels, parts shelves, dollies, supports for heavy components, etc.

5. Tools for meeting: Tables, chairs, white boards, personal computers, etc.

6. Worksheets: Labor-Hour Analysis Comparison forms (A-3page 188?)

COST TEAR-DOWN

Cost Tear-Down is a method by which the costs of competitors' products are compared with those of our comparable products. Because the issue of concern is cost, all cost-related information must be included in the analysis. Information includes component cost comparison, assembly time translated into cost as determined in Dynamic Tear-Down, and the results of other types of Tear-Down discussed later.

The objective of Cost Tear-Down is to assess the total cost to market the product, less general overhead, which is difficult to determine with competitors' products. The purpose of focusing on cost is not to collect and compare total product cost, but to obtain information for improving functions, features, attributes, and identifying their contributions to enhancing customer perceived value. This is achieved by evaluating competitive advantages through analysis, and isolating cost-reduction factors through such analysis.

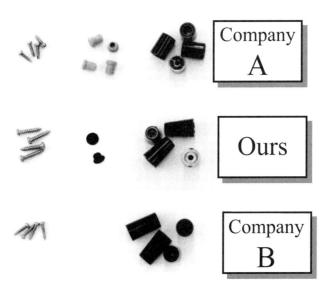

Figure 4-4 Comparison of oven toaster feet

Cost reduction does not by itself improve value. A competitor's product may have a higher cost to increase reliability, maintenance, operations, packaging, etc. A market analysis will determine the competitor's wisdom of increasing product cost to incorporate those features. A value rule to remember is this: If the customer is willing to pay for the features, they have value. Conversely, costly features or functions that the customer is not willing to pay for do not add value.

The information gathered during the Cost Tear-Down process is valuable in planning strategy for improving product functions and customer appeal as well as developing strategies and plans for future product development.

SCOPE OF APPLICATION

Cost Tear-Down is applicable to everything that incurs cost, not only to the physical products, but also to such related items as tooling, packages, accessories, catalogues, distribution, and logistics.

EFFECTS OF COST TEAR-DOWN

In addition to finding ways to reduce the cost of products by comparing cost elements, Cost Tear-Down is also a contributing factor in planning product strategy through the analysis of cost, functions, and quality. In summary, Cost Tear-Down is used for:

1. Determining cost, specifications, and competitive advantages not only for the completed product, but also for the assemblies and parts that make up that product.

2. Finding cost-reducing, function-improving, and value-adding features.

3. Investigating competitors' moves and trends in new products and features.

4. Collecting information for future product ideas.

ESSENTIALS OF COST TEAR-DOWN

The cost information of the Cost Tear-Down team's product is known, or readily available to us. The manufacturing department has process time data convertible to unit, or part cost. Procurement has the cost of items purchased, rather than manufactured.

Because the details of the competitor's manufacturing process and volume are not known, the competitor's cost is usually estimated on the basis of our own, or our supplier's process and volume. Procurement can search catalogs to determine the cost of competitors' purchased items. For custom-produced components, manufacturing can estimate the competitor's cost of purchased items by analyzing their cost to manufacture those components in our own shop.

The differences between our cost and the competitor's are identified on the basis of this comparative analysis. Where the competitor's cost is lower than ours, the team uses the difference as their target for reducing their product cost. Conversely, where the competitor's cost elements are higher than ours, the team can use the cost difference to improve or add value-adding functions and features to our product.

BASIC STEPS

The basic steps of Cost Tear-Down are shown in Figure 4-5

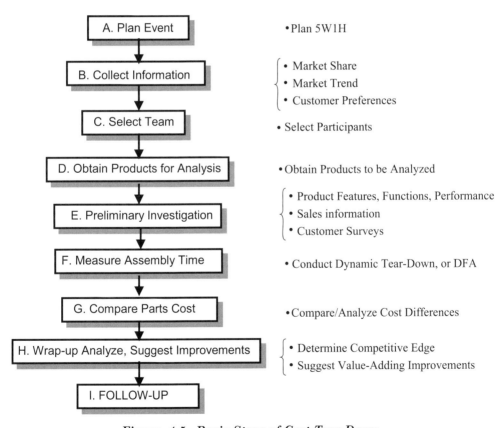

Figure. 4-5   Basic Steps of Cost Tear-Down

SPECIFIC ACTIONS

The description of the basic steps of Cost Tear-Down are similar to the steps described in the Dynamic Tear-Down process.

1. Plan Event

As described for Dynamic Tear-Down, planning is carried out in accordance with the principle of 5W1H: 5Ws – why, what, who, when and where; 1H - how.

2. Collect Information

Information collected can be divided into information from the market-place, and information inherent in the product.

The information from the marketplace includes market share, new product trends, pricing trends, model extensions, accessories, packaging, customer satisfaction, and style changes. Marketing information should be collected from the world market because competitors are not limited to domestic manufacturers or markets.

Product-related information includes advanced design and manufacturing technology, the new use of materials, the ability of materials to be recycled, and impacts on the environment arising from the production and use of the product.

3. Select Members

Product designers and cost analysts are essential members of any Cost Tear-Down team. Participation by personnel from purchasing, manufacturing, and production engineering is also needed in most tear-downs.

To gather the data related to assembly cost, the data accumulated in Dynamic Tear-Down can be translated from time to cost. Participation by process and robotics specialists is most helpful in trying to analyze the cost of assembly if it becomes apparent that the product and assembly process were designed to DFA (Design for Assembly) criteria.

4. Obtain Products for Analysis

Procedures for selecting and collecting competitors' products are basically the same as those for Dynamic Tear-Down. The product that has strong price competitiveness and one which is likely to have some impact in the near future is a good candidate. These products usually come from the competitor that has the highest market share. Products to be collected for analysis also vary with the purposes of the Tear-Down. Is the issue of concern cost reduction? Increase in market share? Making inroads into the global market? All these require change, and require gathering information about the product itself and all related information noted above.

5. Preliminary Investigation

This step also is basically similar to that for Dynamic Tear-Down. Once the products to be analyzed are obtained, the team must prepare, arrange, verify, and adjust the tools, facilities, and schedule. The state and performance of each product must be verified and recorded before disassembly; otherwise, this would be hard to determine once the product is disassembled.

6. Measure Assembly Time

In Cost Tear-Down, the cost of each part of a product is analyzed and compared as it is disassembled. As noted previously, the assembly time of Dynamic Tear-Down is translated into cost.

Dynamic Tear-Down is supported by commercially-available computer software designed for Design For Assembly (DFA) criteria in that it can assess product designs in the same way. Therefore, DFA software can be used in conducting Dynamic Tear-Down studies. This software is very useful in evaluating and improving assembly operations. It allows the Cost Tear-Down team to compare their product with the competitor's using the same evaluation procedure for comparative analysis.

7. Compare Costs by Parts

The basic operation of any Tear-Down procedures is to find differences through comparative analysis. Finding differences through competitive analysis is also the core of the Cost Tear-Down process. The objective of comparing cost by parts is to compare and analyze cost differences in the areas or parts of a product that perform similar functions. The pre-conditions for comparison are as follows:

A. Select the competitor's products that serve the same niche market as the product to be improved. Analyze only one product at a time, rather than multiple candidates.

B. Collect assembly processes available in our company, and any competitive material such as parts lists that would help with the analysis.

    C. The competitor's materials and process cost levels should be comparable with our product costs.

    D. The competitor's amortization and development costs cannot be accurately determined. Estimating such expenses is helpful, but should be excluded as a unit-cost expense item.

Only variable cost is compared in this stage of the process. As mentioned earlier, because the competitor's cost levels and specific processes are not known, competitor's products are examined and analyzed to determine what it would cost to produce their product in our manufacturing facility. It would require an inordinate amount of time to perform a complete bottoms-up cost analysis. The focus of the analysis is not to determine the total operations cost of each part, assembly, or product, but to determine the cost differences between product designs. This simplifies the comparative analysis, reducing the chance of accumulating cost data errors. This approach provides the best way to reach a conclusion within a reasonable time.

To arrive at the cost differences, begin by setting aside those parts that have similar configurations, functions, and cost. Concentrate on the differences in cost for those parts in which our costs differ from those of comparable competitor's parts and calculate the delta, or cost differences, using our cost standards.

Again, at this point in the process, it is only the variable costs that are determined. What is important to determine is:

    A. Whether our product cost exceeds the competitor's product cost

    B. The amount of the cost difference

    C. The cause of the cost difference

It is important to determine the reasons not only for higher cost, but also for lower costs.

8.   *Wrap up, Analyze, Suggest Improvement Ideas*

The results of the analysis will be mixed: our own product less costly than competitor's in some parts and more costly in others. The grand totals represent ours and the competitor's competitive advantage or disadvantage.

A product's competitive edge varies not only with its cost, but also with production volume and its access to the market, that is, sales strategy, advertising, logistics, etc. The objective of Cost Tear-Down, however, is to focus on design and production cost advantage or disadvantage, and those factors that are causing the difference.

After determining the amount and cause of the cost differentials, a review of the parts and other features that should be addressed for improvement ideas are sorted. Those accepted by the team are translated into product and cost improvement proposals. The rejected candidates should include the reasons for rejection. Many ideas are rejected based on faulty information. Justifying rejections allows for validating the reason for rejections. It also allows us to discuss how the ideas that would otherwise be rejected can be accepted. However, it is even more desirable to discuss what we can do to further improve what have been accepted.

The timing for implementation must be discussed, and decisions made, on which proposals should be implemented in the short term, and which proposals are best considered for future (long-term) implementation. Business and product planning, marketing, and finance should be consulted for those performance and value-adding proposals that would incur unit-cost increases to match or exceed the competitor's offerings. Their input will help determine any business advantages as well as how to finance such improvements. These management sources should also be consulted for their endorsement of the cost-reduction product improvement ideas.

The competitive impact of selected proposals, and when that impact will occur, is a major issue in the implementation decision. In projecting the benefits of the improvements over competition, it cannot be assumed that the competitor will not move forward with other improvements of their own. The acceptance of any proposal must be predicated on the improvements having a reasonable competitive advantage life and not be lost with the competitor's next improvements.

Follow-Up

Too often the team involved in the Tear-Down assignment believes that the assignment ends with the generation of product improvement ideas. That event marks the beginning, not the end, of achieving the benefits from the product improvement proposals. The assignment is not complete until the proposals are put into action and the effect of the actions are verified. Benefits are only achieved after the improvement proposals are implemented.

The accepted ideas are put together as a business case within an implementation plan. The plan should also include the name of the selected project

SPECIFICATIONS		Company A	Our Product	Company B
	Power Source	110V AC	110V AC	110V AC
Rating	Rower Rating	810 W	860 W	860 W
	Frequency	50/60 Hz	50/60 Hz	50/60 Hz
Heater Switch		810/590 W	869/430 W	869/430 W
Automatic Temprature Control		Thermostat	Thermostat	Thermostat
Weight		About 2.3 kg	About 2.3 kg	About 2.3 kg
	(Measured)	2.34kg*	2.36 kg*	2.13kg*
	Width	31.8 cm	37.0 cm	36.4 cm
Outside Dimensions	Depth	23.0 cm	22.0 cm	21,3 cm
	Hight	23.1 cm	21.5 cm	21.0 cm
	Width	26.0 cm	26.0 cm	25.8 cm*
Inside Dimensions	Depth	16.0 cm	18.0 cm	17.0 cm*
	Hight	9.0 cm	9.0 cm	8.5 cm*
	Length	22.0 cm	22.0 cm	23.0 cm*
Oven Plate Inside	Width	12.5 cm	12.5 cm	14.0 cm*
Dimensions	Depth	1.0 cm	1.4 cm	1.0 cm*
Cord Length		1.4 m	1.4 m	1.0 m
Warrenty Period		1 year after purchase	1 year after purchase	1 year after purchase
Servise Parts Storage Period		5 years after production stop	5 years after production stop	5 years after production stop

All values based on catalog except marked * - actually measured

FEATURES				
Housing Size		Small in Width	Low Profile	Low Profile
Switch Pannel Position		Under the Door	On Right of Door	On Right of Door
Heater Switch Type		Button Swich	Button Switch	Dial Switch
Heater Switch Indicator		Toase/Pizza	860/430 W	Hot/Mild
Manufacturer's Name		Right upper corner of Switch Panne	Left opper Corner of Door	Left Center of Door
Model Year		Back of Housing	Left of Housing	Top Center of Door
Cautions		Letters & Illustrations on Housing	Letters only on Housing	Letters Only on Housing
Door Handle Position		Top Center of Door	Left Top of Door	Top Center of door
Door Handle Mounting		2 screws	1 screw (2 turning stoppers)	1 screw (2 turning stoppers)
Electric Cord		Long (1.4 m)	Long (1.4 m)	Short (1.0 m)

Table 4-1: Specifications and Features of Oven Toasters

manager, the events scheduled, and major milestones.

EXAMPLES OF APPLICATIONS

Expanding on the product example used in Chapter 3, (see Figure 3-3), the specifications for the oven toasters are shown in Table 4-1.

COST TEAR-DOWN ANALYSIS

Tables 4-2, parts A, B, and C, are the Cost Tear-Down Analysis Sheets. As an example, look at item 8 on part C of this table, which lists cost information for the tray component of the toaster. The tray consists of two parts: the bread tray, and the crumb tray in which crumbs fall and are collected.

Our toaster trays are purchased for 0.45 (see Row 8, column "Price/set"). Competitor A's tray is slightly larger than ours and extrapolation by weight percentage produces an additional 0.0214, making the estimated price 0.4714 per set. Competitor B's cost is estimated in the same manner. Manufacturing methods look similar, so the team decided that the only difference is in the cost of the material.

Examining our crumb tray, note that the tray is smaller than the competitors', as shown in Table 4-2, part (Row 8). Competitor A has a separate piece, which is pressed and paint-coated, then attached to the handle of the tray. Competitor B has an integrated tray and handle, but with a decorative tape of the same color as the cover affixed to the handle.

Competitor A's unit cost is estimated to be 0.47 higher than ours, which consists of 0.37 due to the larger size with better stamping layout, 0.05 for the paint-coated separate handle piece estimated on the basis of our cost table, and 0.05 for assembling the separate piece.

Competitor B's cost increment over ours is estimated to be 0.26, which consists of 0.21 for the crumb tray, 0.03 for the tape, and 0.02 for applying it.

COST TEAR-DOWN REPORT

Table 4-3 shows a Cost Tear-Down Report that covers the overall cost comparison, those factors causing cost differences and suggested future actions. This report example covers the entire oven toaster. Products having a large number of systems, assemblies, and parts, such as a car, are divided into smaller Cost Tear-Down assignments. A Cost Tear-Down report is created for smaller elements and purchased parts such as headlamps and door mirrors. The individual reports are then collected and catalogued; in sum, they repre-

No Components	No	Parts	Qty	Unit price	Our product Cost/set	Our Measured	Our Interesting points	Company A Cost/set	A Delta	A Measured	A Interesting points	Company B Cost/set	B Delta	B Measured	B Interesting points
1 Assembly	1	Cord	1	0.45	0.45	160mm	Place of Stay: 210 mm	0.44	0.01	1540mm	Cord inside housing 5 mm	0.34	-0.11	1175mm	Cord inside housing 0 mm
	2	Plug	1	0.20	0.20		Firm and solid	0.195	-0.50		Easy to insert, good design	0.20	-0.50		Easy to insert, good design
	3	Connector	1	0.10	0.10			0.12	0.02		Harness inserted on side	0.11	0.01		Same as H product; long bend
	4	Protector	1	0.20	0.20		For safety because of switch position	0	-0.2		Not necessary because connector and switch close to each other		-0.20		Not necessary because connector and switch close to each other
	5	Extension Cord	1	0.03	0.03	320mm		0	-0.032		Cord is used (close to switch)		-0.03		Power supply cord used (close to switch)
	6	Terminal A	1	0.01	0.01			0.01	0		Screw- insertion		-0.01		Insert type
	7	Terminal B	1	0.01	0.01			0	-0.01		Soldered (no terminal)	0.01	0.00		Soldered (no terminal)
	8	Joint	1	0.01	0.01		Staked	0	-0.01		Not necessary because no extension cord exists.	0.00	-0.01		Not necessary because no extension cord exists.
	9	Vinyl-wrapped wire clamp	1	0.20	0.20	110mm	Black	0.19	-0.01	102mm	Black, short	0.06	-0.01	102mm	Gray, somewhat short
	10	Assembly	1	0.20	0.20		Staking 3, difference in protector 1, connector 1	0.04	-0.16		Terminal staking 1	0.71	-0.14		Staking 1, connector 1
		Sub Total			1.412							0.71	-1.01		
2 Heater circuit	1	Heater Assembly	2	1.10	2.2		Core recess 11 mm deep	2.04	-0.16		No washer, short porcelain insulator, large heater wire dia.	1.96	-0.24		No washer, few parts, good assembleability
	2	Timer	1	2.50	2.5		Lower reflector used	2.49	-0.01		Core recess 3 mm deep	2.50	0.00		Core recess 11 mm deep (comparable)
	3	Timer Plate	0				Black phenol resin	0.05	0.05		Additional protector plate				None
		Timer Knob	1	0.20	0.20	Weight10g		0.12	0.06	Weight 4k	PP (gray)	0.12	-0.08	Weight 4g	PP (black)
	4	Heater Switch	1	1.30	1.30		Thermostat, fixed with 1 screw	1.05	-0.25		Sheet-type base, small overall size. 1 middle plate eliminated	0.63	-0.67		Different switch structure (turn to switch)
	5	Quick Sensor	1	1.80	1.80			0.63	-1.17		Small; with bracket	0.60	-1.20		Small type
	6	Diode (electronic)	1	0.40	0.440		2 pcs . large	0.8	0.4		2 parts (diode, capacitor)	0.73	0.33		2 parts (diode, capacitor); small
	7	Harness	7	0.02	0.128			0.0123	-0.012		No insulation, short cord, sheet-type harness	0.08	-0.05		Double insulation
		Sheet harness	0				None	0.22	0.22		3 to restrict cord motion				None
		Insulation (cord)	2	0.02	0.036		Woven, 2 pcs ., thin	0.23	0.194		Woven, 2 pcs ., thick	0.01	-0.03		1 vinyl tube
		Plate	0				None	0.05	0.05	Sheet thickness 0.36 mm	For installing electronic parts, 53 by 45	0.07	0.07	Sheet thickness 0.8 mm	For installing electronic parts, 60 by 35
2 pcs . large		Insulation	0				None	0	0		None	0.10	0.10		For insulating electronic plate
	8	Staking (joint)	2	0.02	0.03		2 pcs . large	0.01	-0.02		1 small joint	0.06	0.03		Small, 6
		Protector	1	0.20	0.2			0	0		Not necessary because of switch position	0.00	-0.20		Not necessary because of switch position
	9	Terminal	6	0.01	0.06			0	-0.06		Soldering and bent-over harness	0.01	-0.05		None
	10	Screw (M3)	4	0.01	0.03			0.032	0		4 pcs.	0.00	-0.03		Soldering/staking
	11	Screw (M4)	2	0.01	0.02			0	-0.02		Soldering	0.00	-0.02		Soldering/staking
	12	Assembly		0.21	0.21		Staking solder 3, insulation 3, terminal 6	0.15	0.06		Soldering 4, staking 1	0.12	-0.09		Soldering 2, staking 6
		Sub Total			9.12			8.8703	-1.54			6.99	-2.13		

Tables 4-2A Comparison of Oven Toaster Crumb Trays

No	Components	No	Parts	Q'ty	Unit price	Cost/set	Our product: Measured	Our product: Interesting points	A: Cost/set	A: Delta	A: Measured	Company A: Interesting points	B: Cost/set	B: Delta	B: Measured	Company B: Interesting points
3	Grille assembly	1	Grille	1	0.10	0.10			0.205	0.105		Greater number of wires, additional rings, etc.	0.11	1.00		1 side wire added, good assembleability
		2	Bush	2	0.01	0.02	Length 8		0.021	0.001	Length 9	Somewhat long	1.90	-0.10	Length 7.5	Somewhat short
		3	Spring	2	0.05	0.01	Coil dia 7		0.01	0	Coil dia 7	About equal	0.12	0.02	Coil dia 1	Large coil diameter
		4	Assembly			0.51			0.72	0.21		Greater number of parts	0.46	-0.05		No swaging
		5	Surface treatment (chrome plating)			0.06			0.06	0.015		increase due to greater number of parts	4.50	0.00		About equal
			Sub Total			0.69			1.016	0.331			7.09	0.87		
4	Door Assembly	1	Door Outer	1	0.38	0.38	Sheet thickness 0.45 mm	Length 8 mm	0.43	0.05	Sheet thickness 0.3 mm	Letter printing added	0.45	0.07	Sheet thickness 0.3 mm	2-color printing, no paint coating on back
		2	Door Inner	1	0.34	0.34	Sheet thickness 0.4 mm	Sheet thickness 0.45 mm	0.367	0.027	Sheet thickness 0.3mm		0.39	0.05	Sheet thickness 0.3mm	Bent-sheet metal hinge pin
		3	Hinge Pin	1	0.40	0.40	Thickness 3.2 mm	Inserted hinge	0.03	0.03		2 hinge pins				Bent-sheet metal hinge pin
			Glass						0.36	-0.04	Sheet thickness 3.2mm	Small size: 86 x 240 (166 g)	0.32	-0.08	Sheet thickness 3.2mm	Small size 79x235 (145 g)
		4	Door Handle	1	0.11	0.11	Weight 21 g		0.106	-0.4	Weight 20g	installed on left side, no side tooling, 2 turn stoppers	0.11	0.00	Weight 20g	2 turn stoppers
		5	Tapping Screw	2	1.00	0.02		Chrome-plated screw	0.008	-0.012		1 Zn-plated screw	0.80	-0.01		1 Zn-plated screw
		6	Bracket	2	0.03	0.06	Sheet thickness 0.8 mm	Thin sheet, small size	0.1	0.02	Sheet thickness 1.0mm	Large thick sheet, turn stopper added	0.10	2.00	Sheet thickness 0.8mm	Large thick sheet, turn stopper added
		7	Rivet	2	0.01	0.01			0.01	0		Tubular rivet	1.00	0.00		Tubular rivet
		8	Assembly		0.17	0.17			0.15	-0.02		1 less handle fixing screw, 2 pins added	0.13	-0.02		1 less handle fixing screw
			Sub Total			1.485			1.561	-0.325			1.52	0.03		
5	Reflector	1	Main Refractor	1	0.62	0.62	Sheet thickness 0.3mm	Integral with side	0.68	0.06	Sheet thickness 0.35mm	Separate side, 2 more bends	0.60	-0.02	Sheet thickness 0.3mm	integral with side, good assembleability
		2	Refractor upper	1	0.16	0.16	Sheet thickness 0.3mm		0.19	0.03	Sheet thickness 0.3 mm	Special Zn plating?, large size	0.18	0.02	Sheet thickness 0.3 mm	Somewhat large
		3	Refractor lower	1	0.17	0.17	Sheet thickness 0.3mm		0	-0.17		Not required because switch is on the outside		-0.17		Not required because switch is on the outside
		4	Protector	1					0.12	0.12	Sheet thickness 0.8mm	For protecting lower heater				
		5	Reinforcement						0	0			0.10	0.10	Sheet thickness 0.35 mm	Reinforcement of upper part of reflector, upper
			Sub Total			0.95			0.99	0.04			0.88	-0.07		

Tables 4-2B Comparison of Oven Toaster Crumb Trays

No / Components	Parts No	Parts	Qty	Unit price	Cost/set	Measured	Interesting points	A Cost/set	A Delta	A Measured	A Interesting points	B Cost/set	B Delta	B Measured	B Interesting points
						Our product		Company A				Company B			
6 Switch Panel	1	Panel	1	0.71	0.71	Sheet thickness 0.35mm	3-color silk-screen printing	0.595	-0.113	Sheet thickness 0.35mm	Small, color sheet, 3-color letters	0.57	-0.14	Sheet thickness 0.3mm	Small, no paint coat on back, 3-color letters
	2	Reinforcement	1	0.05	0.05	Sheet thickness 0.35mm		0.08	0.03	Sheet thickness 0.35mm	Small, but bending operation required		-0.05		Not necessary because switch is installed directly
	3	Protect Film	1	0.05	0.05		Necessary because of harness routing	0	-0.05	Not necessary	Not necessary		-0.05		Not necessary
	4	Assembly	1	0.05	0.05			0.03	-0.02	No protector film	No protector film		-0.05		No protector film and reinforcement
		Sub Total			0.858			0.705	-0.153			0.57	-0.29		
7 Cover	1	Cover upper	1	0.78	0.78	Sheet thickness 0.45mm		0.806	0.026	Sheet thickness 0.45mm	Large because of side switch	0.68	-0.10	Sheet thickness 0.35mm	Large because of side switch
	2	Cover side RH	1	0.43	0.43	Sheet thickness 0.45mm		0.426	-0.4	Sheet thickness 0.35mm	Thin-gauge sheet, slits for heat radiation	0.35	-0.08	Sheet thickness 0.35mm	Thin-gauge sheet
	3	Cover lower	1	0.43	0.43	Sheet thickness 0.45mm		0.416	-0.014	Sheet thickness 0.35mm	Thin-gauge sheet, label printed on it	0.35	-0.08	Sheet thickness 0.35mm	Thin-gauge sheet
	4	Cover side LH	1	0.21	0.21	Sheet thickness 0.3mm		0.407	0.197	Sheet thickness 0.45mm	Large size	0.51	0.30	Sheet thickness 0.35mm	Large size, assembling required
	5	Foot	4	0.05	0.2	5 g/pc.	Phenol resin	0.16	-0.04	4 g/pc.	Smaller size, phenol resin	0.16	-0.04	4 g/pc.	Light weight, phenol resin (dull surface)
	6	Foot rubber	2	0.02	0.04	2 pcs.	Rubber	0.08	0.04	4 pcs	With through hole, easy insertion, rubber	0.04	-0.04	0 pcs.	None
	7	Tapping Screw	4	1.00	0.04	M 4 screw		0.032	-0.008	M3 screw	4 small screws used	0.03	-0.01	M3 screw	4 small screws used
	8	Sheet, label	1	0.10	0.10	85×183	One color (black letters on clear base)	0.06	-0.04	48~85	Small, two-color (black and yellow letters on silver base)	0.09	-0.01	60~145	Small, two-color (silver and black letters on clear base)
	9	Model year label	1	0.02	0.15	25×6	Common to other home electric appliances?	0	-0.015	Printed on side cover	Printed on side cover		-0.02		Indicated on sheet, label
		Sub Total			2.36			2.387	-0.254			2.17	-0.08		
8 Tray assembly	1	Bread tray	1	0.45	0.45		Foot screws used for final fixing	0.4714	0.0214		Large size	0.47	0.02		Large size
	2	Crumb tray	1	0.29	0.29		72 positions	0.76	0.47		Large size (+37), 2-piece structure (paint coat 5 + assembly 5)	0.55	0.26		Large size (+21), tape (+3+2), stamping
		Sub Total			0.74			1.2314	0.4914			1.02	0.28		
9 Overall assembly	1	Assembly parts	1	0.09	0.09			0.084	0.084		8 screws, 1 nut, 1 spring washer	0.09	0.09		10 screws, 2 nuts
	2	Assembling, inspection	2		4.5			5	0.5	90 positions	90 positions	4.25	-0.25	66 positions	66 positions
		Sub Total			4.5			5.084	0.089			4.34	-0.02		
Manufacturing cost		Grand total			21.45			21.85	-0.41		(As compared with our product: 98.2%)	18.74	-3.15		(As compared with our product: 85.6%)

Tables 4-2C   Comparison of Oven Toaster Crumb Trays

sent the entire car. The following describes the major topics to be addressed in creating the Cost Tear-Down report.

Conclusion

This section in the report form contains the overall summary of the report. The percentages shown are those of our competitors' (A and B) costs as compared with our cost. The assembly costs analyzed and determined in the Dynamic Tear-Down are included. In addition, differences in major specifications and weights are also included in this column. In the toaster oven example, the product weight is not a significant value-adding characteristic. However, weight in heavy products like automobiles impacts a number of value indicators such as performance, fuel economy, and handling, and therefore requires special attention

Factors Causing Cost Differences

Listed in the "Factors for low" section are the items for which competitors' costs are lower than ours. These items are considered possible sources for product improvement through cost reduction. The next section lists those competitors' cost items that are higher than ours. This report may indicate that our design is better than the competition, or the competitor's products have better performance.

Actions and Requests for Future

This section indicates what actions to take now, as well as recommendations for future actions. The section is the consensus of the full Tear-Down team's suggestions summarized on the basis of all the analyses performed up to this point in time.

Cautions

A. It is not necessary to determine the absolute cost of the competitor's product. Determine the differences in cost, focusing on the reason for those differences. There is nothing gained in evaluating items that are cost equal.

B. Record any use of automation or special tooling used by the competitor to achieve a cost advantage. Determine the cost effect by analyzing the cost to produce the item in our manufacturing facility.

C. Assume the same production volume and cost level as our own product.

Oven Toaster ; Cost Tear Down Report

Date: ___ Japan Electric Appliances Co., Ltd.

Item	Part number / Part description	Weight	Q'ty/unit	Price Unit price / Per set	Analyzed and reported by
	Oven toaster	2.34Kg	1	21.89 / 21.89	

Features (major specification differences, sales points, patents, etc.)

Conclusion	Cost Level		Weight
A Co.	98.4%	Switches on side (range type), button-type heater switch, door handle located left side upper	2.36Kg (+0.02Kg)
B Co.	85.6%	Switches on side (range type), dial-type heater switch, door handle located center upper	2.1 3Kg (-0.2 1Kg)

Factors causing cost differences	COMPANY A (+ -)	COMPANY B (+ -)
Factors for low		
Smaller quick sensor	-117	Smaller quick sensor -120
Sheet type heater switch base, etc.	-25	Dial-type heater switch -67
Shorter harness due to position of connector No protector, no reflector, lower	-69	Shorter harness due to position of connector No protector, no reflector, lower -62
		Simpler heater assembly harnessing -24
Factors for high		
Two electronic parts (diode and capacitor)	+40	Two electronic parts (diode and capacitor) +33
Lower heater protector	+12	Larger cover, lower +30
Larger cover, lower	+20	Bread tray 1.5 cm larger in height +2
Larger crumb tray, 2-piece structure	+26	

Actions and requests for future		Target (difference)
* Smaller quick sensor * Sheet type heater switch base, etc. * Relocate connector. * Simpler heater assembly harness * Shorten cord by 35 cm	* Make bread tray 1.5 cm larger in height (Shorten harness, and eliminate protector and reflector, lower.)	Weight △ 0.25 kg/set Cost △ \2.89/set

Table 4-3   Oven Toaster Cost Tear-Down Report

Cost Tear-Down Tools

Among the tools we use for Cost Tear-Down, the essential ones are the Cost Table and the Price Table. For purchased parts, an Estimate Analysis Sheet is used to project cost differences on the basis of current prices. If, in a simplified analysis method, differences are determined by weight percentage, or area percentage, instruments will be required to determine weight and area.

The results of the Cost Tear-Down analysis are summarized in the Cost Tear-Down Analysis Tally Sheet. The final report is written on the Cost Tear-Down Report form.

SIMPLIFIED COST TEAR-DOWN

The process described above is the conventional way for conducting the Cost Tear-Down procedure. However, evaluating the cost of each part could be time-consuming. Occasionally, the primary focus of the analysis is on competitive differences offered by the competitor's product. The simplified Cost Tear-Down process only addresses those differences and extracts improvement ideas from the differences in each part.

The outline for the simplified Cost Tear-Down process is described below.

Analyzing Procedures

Using the Cost Tear-Down Analysis Tally Sheet, enter the names of the selected components and, if an assembly, the names of the parts that comprise the components. Make note of the observed differences between our product and our competitor's in the space for "Interesting points."

Do not judge the differences at this point in the evaluation. Select those factors that differentiate our product from the competition. Observe and record any quality, dimension and material variances. Record only parts that are different from ours. The parts that have been entered on this form are the items that have some difference between our product and the competitor's.

Evaluating Advantages and Disadvantages

Every difference is compared for cost advantage or disadvantage. Those competitor items having the advantage are marked with a circle. It is again important to note that selecting or rejecting items for evaluation should be avoided at this stage. Such judgment should be suspended until an objective evaluation is performed, as described in the following steps.

Decision of Acceptance or Rejection

Those items circled are selected as prime candidates for improvement changes. However, other items that may not be readily apparent during disassembly may also be candidates for change. These items include, as examples, different grades in bearings, inserts, different fasteners, types of castings, and mating surfaces. A decision must be made at this point to consider theses items with the circled items.

For items that have been chosen, write "A" in the circle for items that can be implemented soon, "B" for those selected to be implemented sometime in the future or in the next model change, and "C" for those that cannot be accepted now.

Evaluation of Effects

For "A" items, determine the cost difference for that part, ignoring at this point the specifics of implementation. Only the unit-cost differences between our product and the competitor's are evaluated.

Summary

The summary of the simplified Cost Tear-Down process is the same as that described in the full Cost Tear-Down Report. Major items that cause differences in cost are listed in the "Factors causing cost differences" section. Both higher and lower cost factors are listed.

MATERIAL TEAR-DOWN

OUTLINE OF MATERIAL TEAR-DOWN

With the growing concern for ecology, more attention is being paid to material. The selection of materials and surface treatment needs to consider the increasing demand for resources conservation and global environmental protection. Materials and surface treatment are therefore items that are receiving more attention in comparative analysis.

Even though coated or heat-treated surfaces look the same, cost may differ depending on coat thickness or pretreatment procedures. In addition, the scrap rates of the materials, or the unused material that is removed to configure the parts, are also analyzed in this Tear-Down process. Material Tear-Down compares and analyzes all the aspects of materials in product design and manufacturing.

SCOPE OF APPLICATION

A. Products that are comparable in function but are different in materials, surface treatment, and heat treatment are included

B. Products that are heavily dependent on material surface treatment to achieve the desired quality, function, and performance of the product and its cost

C. Products with a high ratio of material to labor cost

D. Products with a large variety of materials and surface treatments

E. Products that have high material scrap rates in processing and manufacturing, or that incur high cost for end-of-life recycling or disposal

F. Products that use materials or resources that are likely to become depleted, or are imported

G. Competitor's products that provide a benchmark for materials use, which could be applied to our own products

OBJECTIVES OF MATERIAL TEAR-DOWN

A. Saving materials and labor costs by changing materials, surface treatment, or heat treatment

B. Saving material costs by reducing material scrap rates

C. Saving process cost by selecting materials that are not environmentally sensitive

D. Improving material effectiveness through reducing the variety of materials and surface treatments used in a product

ESSENTIALS OF MATERIAL TEAR-DOWN

Material Tear-Down can be classified into two types depending on its objectives: (1) Comparison of materials, surface treatment, etc., and (2) analysis of material scrap rate. Both directions offer significant opportunities to reduce cost and improve product value.

The costs of materials and surface treatment may not, by themselves, be significant as direct part costs, but they may affect product performance and durability as the product corrodes or deteriorates with use. Lifetime experience impacts customer-perceived value that could influence future product selection and sales. Paint coating, coat thickness, and coating process (electrodeposition, baking, etc.) have significant influence on performance. Heat treatment, hardening method, and case depth have similar effects. Just a fraction of difference in cost may translate into a significant difference in the useful life of the product and its value perception.

The products being analyzed are compared for materials and surface treatment. The section edges of cut samples of the product, or part, are inspected for any difference in materials, surface treatment, or heat treatment. Microscopic metallurgical analysis and other processes determine type and grade of materials, heat treatment, the use of coatings and their thickness, surface finishes, etc., used in the design and manufacture of the product.

The next step in the analysis of material is to determine scrap rate, the costs of all the materials, the relation between the amount of materials fed into production and the net material used in the part. The principle objective of Material Tear-Down is to reduce material cost and improve manufacturing effectiveness. Competitor records are lacking to determine accurately the cost of these materials and process factors. Therefore, the cost to produce the product in our facility, using our cost standards, is used for comparative analysis. When determining the material left after producing the part, the weight or area ratios are often used and converted into material cost.

In many companies the true cost of scrap and unused material is lost, because such costs are accumulated in an overhead account. For example, a stamped part may be made from a relatively inexpensive material. If the part only uses a small percentage of the sheet size, the total cost of the sheet is charged to the part. This information is often lost unless a Material Tear-Down analysis is performed.

The availability of this information will initiate different layouts, or processes to reduce scrap and material part cost.

BASIC STEPS
The basic steps of the Material Tear-Down are shown in Figure 4-6.

SPECIFIC ACTIONS
The follow actions describe the sequential steps for the Material Tear-Down process.

1. Select Theme
The Material Tear-Down theme is usually selected on the basis of the competitor's need to offset competitor advantages. However, Material Tear-Down

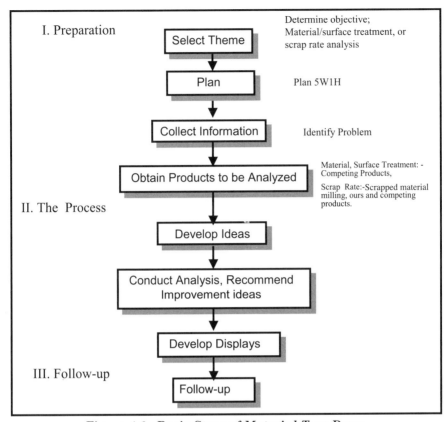

Figure 4-6 Basic Steps of Material Tear-Down

can also be used as an effective way to educate and train employees in design and manufacturing practices. The primary focus of this process can be directed to training employees in good design and production practices by reducing the variety of processes that produce parts, different materials used, surface treatments, and reducing process scrap.

The standards for selecting the competitor's products for comparison are as follows:

Material / Surface Treatment Analysis

A. Competitor and company products with similar functions

B. Competitor and company products with comparable design; those manufactured with comparable manufacturing processes

Scrap Rate Analysis

A. The total amount and variety of materials needed to produce the product, considering both the product itself and its scrapped material

(if possible, include process tools)

    B. Analysis of similar products to detect the extent to which one product is bad or good in comparison with the others

2. Plan

Details of planning 5W1H are similar in many respects to those for the Dynamic Tear-Down. Each item of 5W1H, however, is much different depending on whether the purpose of the Tear-Down is material/surface treatment analysis or scrap rate analysis. The former entails design changes; thus, follow-up actions are the responsibility of the design department. The latter entails process changes; thus, the manufacturing department takes follow-up actions.

The qualifications for the Tear-Down team are as follows:

    A. Responsible designer in charge of the selected theme, authorized to make design decisions

    B. Knowledgeable people with expertise in production and logistical processes, especially with the ability to anticipate potential problems

    C. Cost estimators (from purchasing, cost department or from suppliers)

    D. Material specialists

3.   Collect Information

Many of the costs in Material Tear-Down are not visible to casual inspection. It is especially difficult to determine part cost in companies where cost is collected by cost centers. Under this method of accounting, the operating cost of a process department is amortized by the batch, or lots of parts being processed. The cost of heat treatment, for example, is rarely distributed over the product cost. Even if it is, the distribution is not based on correct weight percentage or volume. Therefore, it is not helpful in determining the true cost of a product.

The same applies to the cost of disposing of waste materials. Usually only rough estimates of the weight and monthly expenses are reported. In many cases, the cost of handling and discarding scrap materials is high, especially so with dangerous materials.

Information on processes also needs to be more exacting. Surface treatment usually requires logistics (i.e., handling and movement to other plant locations). The Tear-Down team must have a clear understanding of the route and conditions under which the parts are moved to the locations of different process steps.

Many potential problems in process and logistics are likely to surface by marking a layout chart or map with routes, then noting the travel and work in process time along which the parts or products go from one step or stage to another.

A review of past field and customer complaints and the history of product changes, as well as a history of similar products, should be part of the Material Tear-Down analysis process. Established standards for corrosion protection or hardness often reflect product complaints in the past. Objective judgment is required to determine if the actions taken then are still valid today, or if technology has advanced and better solutions are now available.

In scrap rate analysis, problems will not be readily apparent. Therefore, no improvement will result unless the processes and the time for processing scrap is considered in addition to time for manufacturing the product. For these reasons, all cost information must represent the true manufacturing cost rather than those costs collected and categorized by the accounting department. Information collected by the accounting department serves a different purpose than data needed to manage a manufacturing facility or data needed to determine either part or product manufacturing cost.

4. Obtain Products to be Analyzed

The material or surface treatments of most competitors' products are process analyzed in Material Tear-Down. In scrap rate and in-house cost analysis, scrapped materials are analyzed in addition to the direct material cost of the products. Scrap rates are dependent on the process selected. Metal stamped parts differ in the volume and cost of scrap when compared to plastics parts, especially when the plastics scrap can be reprocessed as raw material and used again. Also, the material left as scrap differs when comparing metal formed part assemblies to integral precision casting, or parts machined from bar stock.

Material/surface treatment analysis

When the materials and processing are similar, the comparison with competitors' products is directed to material composition and treatment depth (or thickness), which can be determined by microscopic samples, film thickness testers, or surface roughness testers. Surface performance, such as corrosion protection and wear resistance, is also tested using methods to determine the relationship between performance and cost. When approaching the theoretical maximum performance, the large cost to achieve that upper margin of performance is disproportional to the small percentage of performance gained. That is, the cost/performance curve is an exponential relationship.

With the availability of cost performance information, the Value Analysis approach is used to find ways to maintain or improve product performance while reducing cost. A principle of value analysis is that a product feature, such as a material, process, or surface finish has value if the customer is willing to pay for the function that feature offers. Those functions should be identified with the assistance of marketing, sales, and field service.

As an example, the customer doesn't want stainless steel, but the strength, corrosion resistance, and attractiveness offered by the use of stainless steel. After needed or wanted functions are identified, ask, "What performance level and what price are acceptable to the customer?" Will a higher-priced product that will last 100 years attract more customers than a lesser-cost product with a life of 25 years? A floor-covering competitor designed an expensive decorative linoleum tile with a design depth that would stand 50 years of wear. The innovative, more expensive design was less than successful in the marketplace. A post analysis showed that style changes occur about every seven years. Style change was a higher value attribute than the wear quality of the tile.

The VA Tear-Down process, selectively applied to a theme, offers the opportunity to address individual problems and capture opportunities in logistics and surface treatment by understanding the functions served by surface treatment and the performance levels demanded by the markets served.

A way to assess a cost problem is to determine the relation between the weight of a sample section and its cost in terms of unit cost per kg. Multiplying that cost factor by the entire weight of the product, made of the same material and process, will indicate if there is an apparent mismatch between the cost of the product or part and its function. Chapter 6 will discuses this subject in more detail.

There are some limitations to the Tear-Down method in analyzing surface and heat treatment, because the levels of such treatment are often dependent on the available manufacturer's process facilities, rather than requirements of a product. It is not uncommon that a surface treatment higher in grade than necessary is applied to a surface because the manufacturer does not have the process facility to implement the requirements of the function. As an example, a competitor's machine part had as much as a 60µm-thick powder coating, which was in excess of the thickness needed for the function's performance. An analysis showed that the process was chosen because it was the only surface treatment available at the manufacturer's facility.

The above example points to the need to relate the functions needed by the part and process to the performance level wanted by the market.

Scrap Rate Analysis

The objective of Material Tear-Down is to save material cost. Material cost is calculated by the following equation:

Material cost = (Unit material cost x Input material mass)± Scrap disposal cost

Because competitors will not provide such information, it can only be estimated by analyzing the cost of competitor's scrap by assuming they were processed in our facility. The base cost of material can be easily determined by identifying the type of material and process used to produce the part. Each item that makes up the material cost must be considered in performing the detailed analysis. Analyzing the actual products or parts is the first step to identifying problems or product improvement opportunities.

Develop Ideas

A Tear-Down display designed to solicit comments and improvement ideas should contain the information that stimulates such responses. As an example:

Machined products:	Materials, completed products, milling, machining tools, layout in each process, process time
Pressed products:	Completed products and raw material used, scrap generated
Plastics products:	Completed products, tryout runs, sprues, runners, scrap

Visitor comments from machined products display included, "We are making milling chips, rather than a part." " There is more material left on the floor than on the product." Others offered valuable suggestions about the pressed products as to how to reorient the part to get more parts per sheet. A suggestion for the plastics molded part was to reschedule lighter colors first to reduce the number of parts used to purge and to clean the mold before starting the production run.

6.   *Conduct Analysis, Recommend Improvement Ideas*

In addition to many ideas for improvement that are suggested during the processes of information collection and analysis, more ideas are forthcoming as the full product, its process, and scrap are displayed and viewed from different perspectives.

A. Material Tear-Down makes it possible to isolate a function and compare the total of costs of achieving that function's performance. More costly materials and their surface treatment performing the same functions as the competition present an opportunity for cost reduction.

B. Differences in each item compared are translated into suggestions. Selections of ideas as well as various combinations of ideas are encouraged in the quest for product improvement.

C. An issue check list to stimulate suggestions may include, but should not be limited to, the following issues:

a. Can the problems be addressed by ways other than through design changes?

b. Are performance standards higher than needed to satisfactorily achieve the intended function?

c. If factors causing cost differences are not known, can other known areas be improved to eliminate the difference?

d. Can changing the shape result in less costly use of materials and processes?

e. Can reducing the number of mating parts and relieving tight tolerances lower the overall product cost?

Those suggested ideas that can be implemented immediately are sorted from improvement ideas best implemented as part of future actions. As part of the wrap-up, the competitiveness of our product is analyzed by comparing quantitative performance as well as qualitative attractiveness features.

Another important item in the wrap-up step is to highlight trends in material technology. Research on alternative materials has received new momentum as pollution by heavy metals and depletion of natural resources pose potential problems.

7.   *Develop Displays*

The purpose of the Tear-Down display is to focus on those issues that need resolution to improve the competitiveness of our product. How to display the areas of concern and what to display are important considerations because the display design directs observers to develop ideas for those areas emphasized.

No one can determine the case-hardened depth, for example, by just seeing the displayed part. If case depth is an issue to be addressed, a cross-section sample, with enlarged photos and identification of the function served by case hardening, will draw attention to that issue. Consideration should be given to just displaying the parts of the actual product or, if analysis data and other enhancements are needed, by displaying such things as milling, scrapped material, and tools.

The final product must always be displayed because just sections or scrapped material cannot indicate their relation with the completed product. Where several processes are required to produce the product, displaying the output at each step in the process will help observers understand what exactly is done in each process, helping them to detect problems or to offer improvement suggestions.

8. Follow-Up

The procedures are much the same as explained in the section of the Cost Tear-Down. Analysis is not the final objective. Follow-up means an effective use of information obtained as a result of analysis. "Who" and "when" must also be included in the Material Tear-Down follow-up schedule.

EXAMPLES OF ACTUAL APPLICATIONS
Section Analysis

Analysis of a section of sheet metal would be easier to understand if the data as shown in Figure 4-7 is attached.

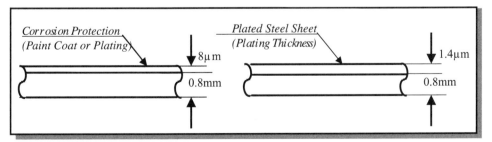

Figure 4-7 Example of Section Analysis of Sheet Metal

Reducing Scrap Rate

Figure 4-8 shows how improving the stamping layout reduced the scrap rate of standard size purchased sheets. In place of small sheets to produce individual parts, a large sheet reduced the process scrap per part by laying-out and nesting different shaped parts on a single sheet.

Cautions

Issues requiring special attention in Material Tear-Down are summarized below:

A. What materials are competitors using? What is the future of those materials (recyclability, resource depletion, etc.)?

B. Are special surface treatments being used? Is the use based on a needed function or performance requirement?

C. How much consideration does competition give to improving performance?

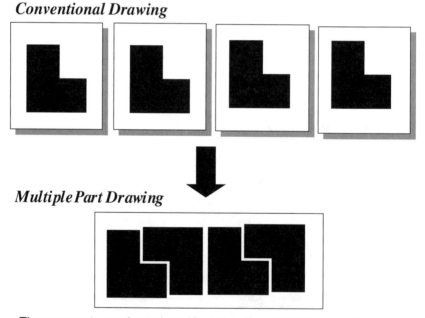

Conventional Drawing

Multiple Part Drawing

The scrap rate can be reduced by improving stamping layout. Sometimes a small dimensional adjustment can drastically reduce scrap rate.

Figure 4-8   Example of Reduced Process Scrap Rated

D. Is the ratio of our material to labor cost higher than normal for the product line?

Tools Used in Material Tear-Down

A. Measuring instruments: Film tester, hardness tester, thickness gauges, micrometer, microscope, gravimeter

B. Work sheets

a. Product comparisons: Materials

b. Product comparisons: Scrap rate

c. Comparison with competitors

MATRIX TEAR-DOWN

OUTLINE OF MATRIX TEAR-DOWN

The types of Tear-Down discussed thus far involve comparison of our products with competitors' products. The Matrix Tear-Down deals with the comparison of products produced by us. Many products manufactured or purchased by the same competitor and facility have a diversity of variations. In Matrix Tear-Down, the focus is to find where common parts across different products can be used to reduce variations and part numbers.

Many part variations in a certain product differ from other product parts that perform the same function. Conversely, other parts with the same function differ because variation is intentional to support a unique feature. For the former case, the intent of Matrix Tear-Down is to try to increase the number of shared parts to improve productivity by larger production lots, reduced investment in part inventory, and reduced fixed costs by shared operations, processes, and facilities. The Matrix Tear-Down analysis identifies base parts (shared parts) and optional parts, which in turn helps improve development efficiency when developing a new product or model.

SCOPE OF APPLICATION

All products, or groups of products and their parts that have similar functions, are candidates for Matrix Tear-Down.

OBJECTIVES OF MATRIX TEAR-DOWN

A. Determine product and part differences to improve productivity through sharing or integration of parts across different products.

B. Reduce development time and cost by reducing the variety of parts and using parts already used by other products (base plus options).

C. Reduce variable and fixed costs.

ESSENTIALS OF MATRIX TEAR-DOWN

A selected product is analyzed to find which part or parts in its assembly are shared with other products (base) and which are unique (option). The shared parts are then compared with all the applicable products within the product line.

The parts performing similar functions are displayed on a matrix table to show which are different from each other and which are the same. The matrix data is then analyzed to determine what problem would occur if parts performing the same function were common. This matrix display helps determine which parts are candidates for common use across products.

BASIC STEPS

The basic steps of the Matrix Tear-Down are shown in Figure 4-9

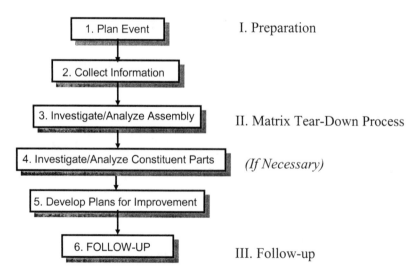

Figure 4-9 Basic Steps of the Matrix Tear-Down

SPECIFIC ACTIONS

1. Plan Event

Selection of Themes

    A. How to select products for Matrix Tear-Down: Usually parts having the same name (e.g., oil filter, front axle shaft)

    B. Standards for selecting the products:

        a. Items having many different types

        b. Items having a wide range of cost variation

        c. Items likely to increase in volume

        d. Items having many different models, produced in small quantities

Selecting Members

    A. Cognizant product designers with the ability to assess changes and the authority to act

    B. Production process specialists; for purchased parts, participation by the supplier's engineers with the necessary information for analysis

    C. Cost analysts with knowledge of part costs and process times (purchasing or cost personnel, parts supplier staff)

    D. Others (if necessary, the product manager in charge of the completed product being analyzed)

Scheduling, Job Sharing

    A. To establish an effective action plan, start with the most effective implementation date and work back.

    B. The Tear-Down team must agree on job sharing. Refer to 5W1H to determine who shall do what, when.

    C. Make sure that all assignments are equitably distributed so that workload will not concentrate on a small number of people. Full team participation results in an appreciation for shared dependencies of different disciplines, higher morale, and a commitment to success.

Major assignments

    A. Leader (overall coordination, preferably a designer of the product to be analyzed)

    B. Secretariat (notice to members, summing up reports, schedule control, etc.)

    C. Finalize the group of parts and collect information on such parts)

   D. Summarize the results of analysis (preferably participation by all
      team members)

Set Targets

   A. Agree on the number or percentage of parts to be reduced as a project
      goal.

   B. Decide on the amount of cost to be reduced (overall saving from
      changes in variable and fixed costs).

2.   Collect Information
Collect Information Unique to the Applicable Parts

   A. Information unique to a part includes performance, cost, the final
      product, past problems relating to the part, time when it was
      designed, history of production volume, current volume, and
      quantities supplied as service parts.

   B. Display drawings as well as actual products.

Collect Information on Competitors

Gather all available competitor information, including service and repair
manuals, part lists, test results, market surveys, and the competitor's actual
product.

Determine Market Needs

The market analysis should include, in addition to user responses, the
dynamics of the market, potential environmental concerns, emerging technology that relates to the product, and future market style trends.

MATRIX TEAR-DOWN PROCESS

   1. Investigate/Analyze Assembly

   2. Investigate/Analyze Constituent Parts

3.   Create Matrix Chart

Each product model is different in some respect. The matrix chart is used
to identify those differences. (Refer to the standard form in the appendix.)

   A. List all the derivative products that have been developed from the
      base product (Variations are developed from an earlier-developed
      base product. Comparison and analysis are easier when variations
      that share a base product configuration are listed together.)

   B. Record the history for each derivative and how it differs from the
      base product.

C. Enter each part of each assembly and the product to which this assembly belongs in the matrix chart.

4. *Analyze*

A. Select and sort parts and assemblies that are shared among products and those that are not.

B. When cost and output must be evaluated in the analysis process, compare by part weight, capacity, performance, or cost to identify problems or improvement opportunities.

C. Compare and analyze major parts that appear to affect cost or cause the growth of variations.

D. To reduce the number of parts, select those units that have disproportionately large numbers of parts in their assembly.

E. Analyze both reducing the number of parts and using common parts across product models in order to reduce the number of assemblies.

F. Use the same procedures for both parts analysis and assemblies analysis.

G. Enter each part and its relation with the applicable assemblies in the matrix chart.

5. *Develop Plans for Improvement*

A. With an understanding of part differences that perform similar functions, solicit product improvements to resolve the differences.

B. Which assembly or component parts can be modified to achieve commonality?

C. Is a change to achieve a common part contributing to product improvement goals? Can an existing part be used and modified to reduce the cost of the different part?

D. Would changing the base part to a new configuration be more effective than adapting the base part to the new design?

E. Can high cost factors be eliminated or replaced with lesser cost parts or assemblies that perform the same functions at an accepted quality level?

F. For purchased items, are there parts that have been produced for other products that can be used across product lines?

OIL FILTER - Matrix Tear-Down Analysis Form

Base Process Part No.	9211-4042-0
Base process	Oil Filter Assembly
No. of use per unit	
No. of Processes — Max 13, Min 10, Average 11	
Date	
Division XXXXXXX Yshihiko Sato	
No. 9211-4042-0 Oil Filter Assembly	

Analysis

Number	Process Name/No. Process description	Manual	Automatic	Walking	4042-0	4053-0	4080-0	4090-0	4130-0	4122-0	4100-0	4111-0	4140-0	Remarks
	No. of product units made/ month				5000	2000	2000	700	300	5800	900	500	10	
1	Air blow — Eliminate foreign matters from body inside	7	15	3	●	●	●	●	●	●	●	●	●	
2	Spill-valve mounting — Spill valve mounted & plug-tightened	9	—	2	●	●	●	●	●	●	●	●	●	
3	Pressure valve mounting — Pressure-adjusting valve mounted & plugtightened	10	—	2	●	●	●	●						
4	Bypath valve mounting — Bypath valve mounted & plug-tightened	8	—	2	●		●	●						
5	Hydraulic gauge plug — Plug mounted & tightened with seal-tape	15	—	4		●								
6	Blind plug mounting — Plug & washer mounted at the oil tap screw section & tightened	14	—	2	●	●	●	●	●	●	●	●	●	
7	Drain plug mounting — Oring inserted, center bolt at the body, and drain plug tightened	13	—	0	●	●	●	●	●	●	●	●	●	
8	Element washer — Element-setting spring & washer mounted	9	—	3	●	●	●	●	●	●	●	●	●	
9	Element mounting — Element mounted onto case, center bolt	5	—	2	●	●	●	●	●	●	●	●	●	
10	Case Oring mounting — Oring mounted on the grooved area	7	—	2	●	●	●	●	●	●	●	●	●	
11	Center bolt tightening — Assemble the body with the filter section	4	5	2	●	●					●	●	●	
12	Blind plug tightening — Plug tightened to the fabricated hole area	7	—	2	●	●					●	●	●	
13	Precision test → air blow — Water tightness test, dripping & stain eliminated, date of mfg.	10	—	4	●	●	●	●	●	●	●	●	●	
14	Packaging — Vinyl bag packing	5	—	1	●	●	●	●	●	●	●	●	●	
	Manhours & Total No. of Processes	123	20	31	11	11	12	12	9	9	9	10	9	
					Difference from spec	with hydraulic gauge	Bypath yes	Bypath yes	Single function type	Single function type	Single function type	Single function type	Single function type	Concept for futue: Toward Cartridge-equipped models.

Sketch (Scanner) Body (ADC12) Filter — Spill valve — Basic spec.

Table 4-4   Example of Oil Filter Matrix Tear-Down

G. What cost advantage can we achieve?

H. For future actions, record those factors that prevent common use of parts. Try to overcome those factors in new design development.

I. Sort out and select improvement suggestions.

J. Display the results of the study.

Cautions

A. All parts made common do not necessarily result in cost reduction. Selecting a more expensive part design as a common part may add cost to some low-cost models. Each case supporting commonality must be justified.

B. Cost advantage resulting from commonality is assessed using the following relationship: the cost to implement the change divided by the (delta) unit cost advantage due to commonality. This determines the break-even point (BEP) in units, for the proposed change. If the unit cost change to achieve commonalty were greater than the current approach, there would be a cost penalty, not a break-even point.

$$\text{BEP} = \frac{\text{Implementation Expense}}{(\text{Current Unit Cost} - \text{Proposed Unit Cost})}$$

C. Evaluation varies with the stage of product development (planning, production, service), difference in processes, and sizes of parts.

6. Follow-Up

A. Improvement suggestions that have survived to this point are summarized on the Tear-Down Suggestion Sheets for further action.

B. Suggestions that have been set aside, but provide some possibility of acceptance in the future, are compiled as VA suggestions and sent to pertinent design departments. There, they are stored in an idea bank.

Examples of Applications

Table 4-4 summarizes the result of the Matrix Tear-Down conducted on an automotive oil filter.

Cautions

The following points require special attention in Matrix Tear-Down:

A. Conduct a team meeting to explore all problems that would occur as a result of commonality.

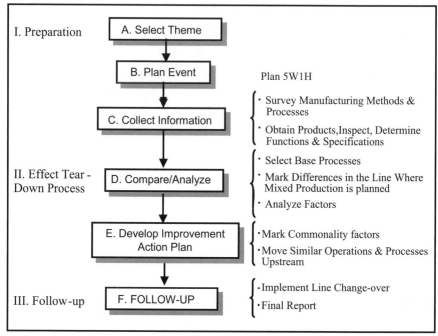

Figure 4-10   Basic Steps of Process Tear-Down

B. The matrix is a standards form of analysis. The matrix can be used as a standard process, but how, and to what the matrix is applied, determines the form design.

C. Tools Used for Matrix Tear-Down

D. Worksheet: Matrix Tear-Down Report (Assembly) (See appendix)

E. Worksheet: Matrix Tear-Down Report (Part)

PROCESS TEAR-DOWN

OUTLINE OF PROCESS TEAR-DOWN

Diversification has produced a great number of different products and product models to satisfy a broad and diverse range of functions and features wanted by the market's customer base. Many products manufactured in relatively small quantities are similar, but different in details. This trend toward customized products is likely to expand, posing problems for the manufacturing industry. A solution to this problem is to manufacture products with common processes on the same production line.

An example is today's telephone. The base model contains a transmitter, a receiver, a ringer, and a way to direct voice signals. This is the base model. However, today's telephone contains a large proliferation of optional features providing special functions and a great variety of product offerings.

Process Tear-Down is similar to Matrix Tear-Down in comparative product analysis, except that the emphasis is on process. Many similar parts or products are manufactured using similar processes. The Process Tear-Down procedures compare and analyze the details of all of these processes, from the initial to final process, for the purpose of integrating similar processes, modifying each process to deal with production of different products on the same line, and improving line effectiveness significantly. A result of Process Tear-Down is a production line that manufactures small-volume parts efficiently at low cost.

An effective way to effect product improvement and cost reduction is first to develop as many common parts as practical. The remaining unique parts are then analyzed using Process Tear-Down to determine which of those parts can share the same process line.

SCOPE OF APPLICATION

    A. Products having similar manufacturing or assembling processes

    B. Products manufactured on different production lines because some functions have been added or deleted from their base products

    C. Those having disproportionately high cost penalties in comparison with their benefits

OBJECTIVES OF PROCESS TEAR-DOWN

    A. Commonality, standardization, and process simplification

    B. Accelerating commonality of products (a product = base product + optional parts)

    C. Combining like process facilities, saving space by producing more than one product on the same line

    D. Reducing product processing and assembly costs as the result of achieving the above objectives; additionally, reducing the investment in new process lines.

ESSENTIALS OF PROCESS TEAR-DOWN

The first step in the Process Tear-Down is to record the current process of each part on a matrix chart, which enables us to quickly see process differences.

Next, product processes that are similar to each other are moved upstream in the process sequence whenever possible, and those unique to individual products are moved downstream. For upstream processes, the team studies ways to share the same machines, the same facilities, and the same tooling by modifying the products or the methods wherever possible. This establishes a process center that can accommodate a variety of products sharing the same process. If sufficient quantities are projected, the additional benefits of automated production can be studied for this stage. The larger the quantity of products that share a production line, the greater the opportunity for expanding shared production lines or adding products to shared production lines.

BASIC STEPS
The basic steps of the Process Tear-Down are shown in Figure 4-10.

SPECIFIC ACTIONS
As with other types of Tear-Down, selecting the Process Tear-Down theme is usually based on the company's need to reduce cost or improve a specific product. However, when Process Tear-Down is selected for education and training purposes, the processes studied encompass several product groups. The following points should be noted when selecting the theme, that is, the processes to be worked on.

1.   Select Themes
First, select the base processes, that is, the processes where the manufactured products can serve as the standard. The guide for selecting the standard is as follows:

A. Processes with which a relatively large quantity of products are manufactured

B. The processes producing products requiring a large number of subsequent processes

C. The processes requiring large processing or assembly labor-hours

Product group candidates for mixed production or commonality include:

A. Product groups that are manufactured by identical or similar methods or processes

B. Product groups that are manufactured in separate or different methods or processes, but have the same or similar functions

C. Product groups that have a large number of parts because of diversity in applications, and for which there is a strong need for commonality

2. Plan Event

The base for planning is 5W1H in this Tear-Down as well. However, "what" is already found in the preceding step. The following attention must be made to address the remaining items.

Select Team Members

A. Person responsible for control in the applicable processes

B. People with expertise on the functions of the products manufactured in the applicable processes (design and test personnel, etc.)

C. People who can evaluate man-hours (cost experts, etc.)

D. Process design personnel for the applicable processes who have authorization for quick decision for change

Scheduling

As with the previous Tear-Down applications, action plans should be based on the most effective implementation date, then work back.

Set Targets

Targets are established to support the goals of the business plan and include issues such as how many assembling lines to eliminate, the percent of parts commonality, and labor hour reduction.

Job Sharing

A. All team members are assigned supporting roles.

B. Major jobs

Leader: Overall coordination, orientation, adjustment among members

Recorder: Notice to members, report summaries, schedule control

Process analysts: Help team comparing company processes with competitor's processes

All team members: Suggest improvement ideas

3. Collect Information

Process Tear-Down addresses products and processes in our company, rather than the competitors' offerings. Therefore, quantitative and historic data is readily available. Because the objectives and project performance have been established before the data collection process, potential problems and emerging improvement ideas begin as information is collected. When collecting information, we should be sensitive to the rejection and repair frequency of a process and those process elements requiring specialty skilled personnel.

The information to be collected is as follows

A. Process charts and standard times for machine and manual operations

B. Process layout charts and personnel placement

C. QC Process Charts, operation standards sheets, and conditions (restrictions, regulatory, and environmental requirements)

D. Tooling, facilities, and equipment used in each process

E. Production capacity and records of each process

Some cautionary notes when collecting information:

A. Production methods and processes should be expressed in operational metrics.

B. Cost data should display costs incurred by each process.

C. Information on parts that are used in multiple processes must also be collected (functions, specifications, sources, prices, commonality, or relationships with other products, etc.).

4. *Compare/Analyze, The Core of Process Tear-Down*

A. First, determine the standard process for the base product. The description of the standard process includes parts assembled in each incremental process, operation procedures, details of operations, time, number of line operators, tools and facilities, and capacity.

B. Select those candidate processes of the product groups that are to be compared and analyzed for potential mixed production or commonality. The details of required information are the same as those for the standard processes.

C. Enter the collected information on the matrix.

D. Highlight differences and determine if commonality can be achieved by changing the process sequence or by modifying the process.

E. If the team concludes that the processes chosen have too few similarities, Process Tear-Down may not be the best theme. The following procedure is therefore suggested:

F. Analyze the process for each product.
   a. For each product, make a card containing process information, with a different color for each product.

b. Enter such information as serial number or process designation, parts assembled in that process, details of operation, and process time

c. Arrange the cards to indicate similar operations among products. If there are very few or no common processes, determine what can be done to the processes and the parts to achieve commonality.

d. Rearrange the order of process cards so that common processes are upstream and unique processes downstream. Sorting the process cards in this manner will suggest potential areas for improvement.

B. Factor analysis: Conduct factor analysis for each item that indicates difference from one process to another.

Why is the order of processes different from that of the base process?

Why does the process have an operation additional to the base process?

Why does this process require highly skilled workers?

C. If any process differs from the base process and has factors that incur higher cost, analyze that process with focus on:

Differences in incremental processes and differences in specifications

Functions for upstream and downstream processes that require high ly-skilled workers

5. Develop Improvement Action Plan

The success for Process Tear-Down improvements depends on the ability to achieve process commonality. Most of the benefits of the Process Tear-Down occur when more than one product group can be produced on a common process line. Implementation of process improvement proposals should focus on the following:

A. Arrange the manufacturing line so that processes which are common to different products are placed forward (upstream) on the line and processes unique to each product are placed later (downstream) on the line. An idea to consider is to produce subassemblies before feeding the products into the line.

B. Determine potential problems in rearranging the process line and develop contingency plans.

C. Consider converting common processes into unit modules or process centers, then separate upstream or downstream processes that are not common.

D. With an increase of parts and products through a common process, consider the economics of automating the process line (automated loading and unloading, automatic machining and processing, common tooling, etc.).

Improve Processes with Potential Problems

A. Use research lessons learned at other process lines to resolve problems of frequent repairs or low flow-through rates. Sub-divide those operations and simplify each segment. If necessary, consider changing the product design to accommodate the common or improved process

B. Determine how to change products so that their processes, facilities, tooling, etc., can be common. Although changing the product design is not the objective, determine which design characteristics block commonality efforts. Also determine the impact on design performance from changing the design to accommodate common processes.

C. If kits or parts prepackaged for a process line do not match the proposed process, try dividing the kit into several kits to match the ideal process layout. This will change the composition of the assembly kit in which the parts are being delivered to the process.

Find Optimum Cost

A. All proposals for common processes must be cost justified.

B. In conducting a business case justification, determine the investment required to implement the individual or groups of proposals. Productivity and flexibility improvements, as well as changes in production volume, are factors to be considered when developing a business case.

C. DFA, (Design for Assembly) explained earlier, is a good companion discipline for ProcessTear-Down. By redesigning the part orientation and types of fasteners used in product assembly, the goal of a fully-automated process is achievable.

Cautions

A. Implementation effort and potential benefits are not always proportional. Look for ideas that require little change, but produce big improvements.

B. Investment must be considered in relationship to the product's future life (change in volume, product features).

C. Where major process changes are involved, consider setting up a pilot process to uncover any potential hidden problems, then verify the potential benefits. A pilot process can also be used to train operators.

D. If any change has been made in the product design, make sure that

Oil Filter: Matrix Tear Down Report (Assembly)

Part number | Date | company Reporter

Item No.	Part No.	Base item No.	Yr./Mo. of Market Introduction	Filtering space (m²)	Flow qty -litre/min.	Mounting Method	Drain? Y/N	Oil cooler? Y/N	Bypath filter Y/N	Filtering space of the bypath filter	Other factors (comments)	Weight (Kg)	Weight change +/-	Cost index (%)	Total output – No./month
1	9211-4042-0		00/3	0.5	57	←						5.4	−		5000
2	9211-4053-0	1	03/8	0.5	65	Cartridge	Y				with a plug for hydraulic gauge u	5.5	−	0.93	2000
3	9211-4080-0	3	98/2	0.9	84	Cartridge		Y			oil-cooler bypath valve added	8.5	+	0.96	2000
4	9211-4090-0	4	02/2	0.9	75	Cartridge	Y	Y			plug & washer added	8.1	+	1.05	700
5	9211-4130-0	5	03/5	0.9	75	Cartridge		Y	Y			8.4	+	1.04	300
6	9211-4122-0	7	02/9	0.7	80	→			Y	0.35	without heli-sert for the oil-tap screw portion (ADC) somewhat different shape	6.9	0	1.05	5800
7	9211-4100-0	7	98/7	0.7	80	Center Bolt			Y	0.37		7.2	+	1.00	900
8	9211-4111-0	8	00/9	0.5	65	Center Bolt						6.2	+	1.11	500
9	9211-4140-0	8	02/4	0.5	65	Center Bolt			Y		without a case nor mounting marks	6.2	+	1.65	10

Output Quantity: Max / Min / Average use per unit — /units
No. of types / Total investment / Average use per unit — /units
Additional investment

Adaptable car model — Output qty. by models

Item No.	C790Eng KED51 (5000)	C240Eng KED52 (2000)	C240Eng TKD23 (700)	(4PA1Eng) TKD24 (1500)	(4PA1Eng) TKD54 (1700)	(4PC1Eng) TKD55 (1900)	(4PB1Eng) KS21 (710)	C330Eng (1400)	C421Eng (1500)	C440PYEng (500)	C440PWEng (300)
1	●										
2									●	●	
3		●									
4											●
5											●
6			●	●	●	●	●				
7								●			
8							●	●			
9											

Futuristic comment & request:

* As long as all TK model cars in 4P series are using common parts, actions for cost reduction purpose are considered feasible throughout all models.
* As for C790 and C440, the engine-mounting surface is same. So, it seems to be possible to combine the two into one under the objective of "using the same cartridge as used in 4P series, PROVIDED that a compromise plan regarding the issue of interfered"

As for the filter use for power machines, it should be ideal if the same filter as for cars be used as it is.

Table 4-5   Results of Oil Filter Process Tear-Down

its basic functions or principle reasons for the existing product are not adversely affected.

6. *Follow-Up*

Prototype products that incorporate major improvement and process changes may not reflect the production model. Prototype lab models are generally hand produced rather than made with production tooling on production lines. The product's performance can be accurately assessed with prototype models. However, without production tools and equipment, process problems can only be assessed and analyzed based on past experience and simulation. The reason for not using production tooling or production processes to build prototypes is the cost of the investment. Therefore, the best that simulation and past experience can achieve is to reduce, but not eliminate, the degree of uncertainty associated with the economic risk of implementing process improvement changes.

PRE-PRODUCTION PROTOTYPE

Although the prototype model may not duplicate the production version, care and thought is given to the process that will produce the product in production.

A. Plan the lines, modifying them to include the product's production volume and expected life cycle.

B. Lay out and locate where production equipment, tooling, jigs, etc., that will be used in each improved process will be placed.

C. Try out actual processing or assembling to verify the effects of the changes. Small-volume production must also be verified on the line. Simulation or engineering studies alone could be misleading.

D. Validate the expected process time improvement. If the process time does not meet expectations, taking into account the operator's learning curve, determine and correct the cause.

TRANSFER TO PRODUCTION

Move on to production if pre-production run proves satisfactory. If not, make further improvement adjustments until improvement goal is achieved.

Audit the process production line for quality before running full production schedules.

FINAL REPORT

Once the mixed production line is complete and products begin to roll out, that fact alone is the best thing to report.

The best closing report is the successful installation and process flow that returns the expected benefits. The report appendix should contain the team's activity records and details of achievement.

AN EXAMPLE OF APPLICATION

Table 4-5 shows an example of the Process Tear-Down study for an automotive oil filter. This analysis indicates that, if processes 3 through 6 are moved to just prior to inspection, all the remaining processes can be shared with other parts. Subsequent discussion on this point then revealed the possibility of integrating parts assembled off line into the processes. In addition, the section "Futuristic comments & requests" has an idea for reducing the number of processes by changing the unit into a cartridge.

Cautions

    A. Are there multiple uses by the production line facilities, equipment, and tools? If not, what can be done to achieve this objective?

    B. Which parts of the product or process can be changed to enable multiple process application?

    C. What are the future plans for the line? Is the line obsolete, or can it be updated?

Tools Used in Process Tear-Down

The worksheet, "Process Tear-Down Report," appears in the appendix. However, the report format is only a guide. The forms themselves should be modified to best fit the company, product, and needs of the VA Tear-Down team project.

STATIC TEAR-DOWN

Static Tear-Down is the first in a growing number of applications for the Tear-Down process. This original Tear-Down process, used for comparison analysis by disassembling and displaying the product, was conducted as a forum. The purpose of the display was to stimulate product improvement ideas. The VA Tear-Down process took this concept further by integrating Value Analysis principles, developing improvement strategies, and coordinating all follow-up and implementation actions.

There are two major steps to Static Tear-Down. In the first step, the selected products, parts, or materials subjected to Dynamic Tear-Down (ours and the competitor's) are displayed as they are disassembled into their component parts. The team then reviews and decides how to analyze each part.

In the second step, the ideas for improvement that were suggested as a result of competitive analysis are used to form product strategies. Seeking improvement suggestions may also be done in the first step, but it can be done more effectively in the second step, especially if the ideas proposed earlier are displayed in this step with the relevant data. Displaying these ideas that were developed in the first step will now stimulate new ideas and combinations of ideas while reducing duplicate suggestions.

SCOPE OF APPLICATION

As mentioned above, the static Tear-Down is the origin of the VA Tear-Down methodology. The process technique is to analyze by comparison. Static Tear-Down is not limited to hardware. Anything that can be identified by its function and displayed is a Tear-Down candidate. Some examples of Static Tear-Down include book production, logistic routing charts, process personnel placement charts, tools, fabric mills, automobiles, and automotive components.

OBJECTIVES OF STATIC TEAR-DOWN

The objectives of Static Tear-Down include the following:

A. Generate improvement ideas from a wide range of disciplines.

B. Rate the overall performance of our products and processes by comparing them with competitors' offerings.

C. Develop a team culture of shared problems, resolutions, and a better understanding about our competitive status in the markets served.

ESSENTIALS OF STATIC TEAR-DOWN

A. The display must be designed so that it can be understood at first glance. The layout must be such that visitors can readily identify how and where the products or parts displayed differ from each other. The display should highlight their features and functions, and how our product compares with the competition.

B. The display must have visitor appeal so that visitors enjoy coming and bringing associates. The environment of the Tear-Down room and the themes must be arranged to interest many people with diverse disciplines, especially management-level executives. This arrangement will broaden the range of ideas and gain management endorsement and sponsorship.

C. Collect fresh, innovative ideas. Sort and cluster ideas and develop the improvement proposal.

BASIC STEPS

The basic steps of the Static Tear-Down procedure are shown in Figure 4-11

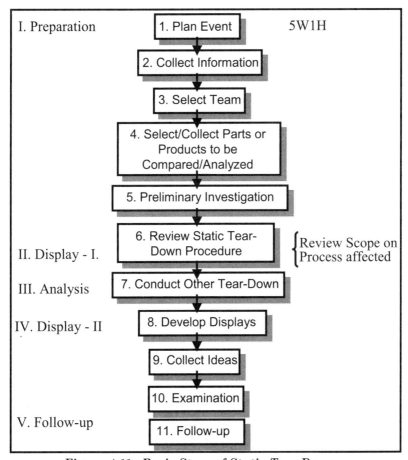

Figure 4-11 Basic Steps of Static Tear-Down

SPECIFIC ACTIONS

The following actions identify the steps of the Static Tear-Down.

1. Plan Event

A. As in previous Tear-Down themes, the planning base is 5W1H.

B. Specify the objective of the Tear-Down project. Is the primary objective to overcome competition advantages, develop new products and processes, train personnel? The Tear-Down theme and approach varies with the project's objective.

C. Define the scope of application; should it cover all parts or only the parts to be changed?. Should other in-house products with essentially the same function be included? The scope of the analysis is a factor in selecting competitors' products for comparison.

D. Which type or types of Tear-Down will be selected? Only the Static Tear-Down or all the types of Tear-Down starting from the Dynamic Tear-Down? Where multiple Tear-Down procedures are used, Static Tear-Down serves as the concluding Tear-Down process.

E. The most effective end date for the Tear-Down project is an important early planning issue. Although the results of the project may promise significant improvements, either large procurement commitments or a new scheduled product introduction date can have priority. The schedule of the Tear-Down plan must begin with determining the implementation date and then working back to the start date.

F. Set targets. If the primary objective is cost reduction, set a target amount for the team. The target should be aggressive, but achievable. The cost-reduction target should also be credible, based on competitor analysis or the needs of the business plan. With the target(s) established, we can discuss and resolve strategies for achieving the cost reduction goal.

2. Collect Information

A. Information is needed for a full understanding of the situation in which the product to be analyzed exists. Which is the target competitor: the market leader or the direct product challenger? Is the market leader's product also the lowest price? How competitive is our product (price, marketability)? Does our product reflect the latest technology? Are sales and distribution issues to be considered?

B. What competitor features and functions are attractive to the market?

C. Are there potential changes in marketing direction that would affect the life cycle of the product?

3. Select Team

A. Tear-Down team members must be well qualified in their field.

B. Included on the team:

a. Product designers with the ability and authority to make product improvement changes. Cost analysts or estimators familiar with process and purchased costs. Previous Tear-Down team members who have any Tear-Down experience with the product being studied.

b. Project team members must have assigned activities and responsibilities. Those assignments include the team leader, recorder, and personnel to disassemble and display the product, coordinate with outside sources, follow-up, etc.

4. Select/Collect Parts or Products to be Compared/Analyzed

A. The strategy and approach selected influences which competitors' products to select, what to compare or analyze, and at what stage to discover problems and develop ideas.

B. If the objective is cost reduction, the price leaders in the market must be selected. If the objective is to develop a new product, products incorporating new technology must be selected.

C, Those products most popular must be selected.

D. If the objective of the Tear-Down is commonality or reduction of the number of parts, displaying all comparable products in the product line is recommended.

E. Displaying some past examples of successful Tear-Down may encourage participants to offer improvement suggestions.

F. Also consider in this step how best to display products and parts to highlight the differences.

5. Preliminary Investigation

Verify the location and size of the Tear-Down room and the disassembly space. Also verify facilities, disassembly equipment, tools, material storage, display equipment, etc.

Determine if anything is missing, when it is needed, and how to acquire the missing items.

6. Review Static Tear-Down Procedure

A. If Static Tear-Down is part of another team's Tear-Down process, the products are in the disassembled state. For example, if Dynamic Tear Down was previously completed, the product is already disassembled and the analysis performed.

B. During planning, in the preparation stage, the task ahead has been outlined. The team observes the products again through different eyes and determines what additional analysis would be effective in conducting the Static Tear-Down.

C. The results of that discussion are added to the plan.

7. *Conduct Other Tear-Down*

The VA Tear-Down calls for conducting various types of analysis as necessary to achieve the specific goals of the theme. The team has already decided what to analyze as described in steps above. The purpose of Tear Down is not analysis. The purpose of Tear-Down is the development of ideas through brainstorming and soliciting suggestions, using all information available, finding opportunities, and successfully implementing the improvement proposal.

8. *Develop Displays*

A. The objective of this display is to raise ideas and report the results of analyses performed. The method of display must match those objectives of the project.

B. The main points of the display are: "Make differences easier to understand," "Don't make the visitors read, let them see," and "Express quantitatively whenever possible."

C. Who do you invite and when do you want to invite them? When do you send invitations? Who should be visitor guides? Visits by senior

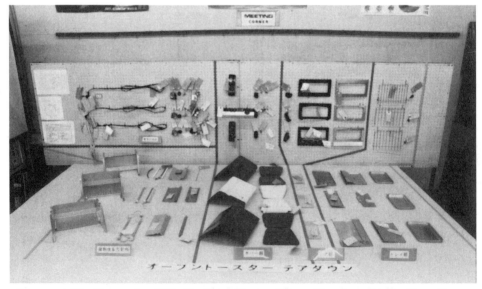

Figure 4-12 Example of Static Tear-Down - Oven Toasters Static Tear-Down as applied to oven toasters.

managers imply interest and endorsement; they motivate others to visit the displays.

D. Visits by employees offer the opportunity for education. Do not exclude anyone from visiting the displays.

E. Invite suppliers to see how their products are being used. A better understanding of the role of their product will result in better suggestions for improvement.

9. Collect Ideas

A. About three-fourths of improvement ideas occur prior to or at step 5 under Develop Displays. Step 5 is designed to stimulate improvement ideas. However, collecting ideas does not stop there.

B. When many and diverse visitors view the displays, their ideas extend over a wide range. Their comments and ideas are recorded on an "Idea Memo" or "Tear-Down Suggestion Sheet" available at the display area.

C. Displaying the ideas as they are suggested is a good way to improve the quality of those ideas. It allows for expanding or combining previously submitted ideas and avoiding duplications.

10. Examination

A. Examination is a two-step process. First, the Tear-Down team members make a preliminary examination to group the ideas roughly into those which are acceptable and those which are not. Next, the specialists affected by the proposed ideas are invited to make a final decision about how best to implement the proposal.

B. Examination is conducted in the Tear-Down Room in front of the actual products displayed.

C. Differences in the competitors' products should be analyzed to determine if the differences result in improved performance, lower cost, customer appeal, or other competitive values.

11. Follow-Up

Unlike other Tear-Down themes, Static Tear-Down appeals to a broad range of interested people. It is, therefore, important to respond to all the suggestions submitted. In addition to acknowledging each idea, when the idea progresses through the acceptance, test, and prototype stages, the status should be posted on the display and the author of the idea credited. This will stimulate communications, encourage people to re-visit the displays, and motivate more people to participate in offering improvement suggestions.

EXAMPLE OF ACTUAL APPLICATION

DISPLAY PROCEDURES

Tools Used in Static Tear-Down

No tools unique to Static Tear-Down are necessary. A variety of tools can be used to accomplish the Tear-Down process and make the display attractive and easy to understand. Some of the tools and equipment recommended are:

A. Tables. Meeting room tables or working benches can be used. Use flat pallets for displaying heavy objects. An isolation screen should be placed around a display to eliminate visual distractions and focus attention on the display.

B. Boards. Perforated boards as shown in Figure 4-13 or mesh boards can be used effectively to display small objects. Reports and sketches, or data can also be placed on the board to aid in understanding the functions of the items.

C. Hooks. The objects must be displayed in such a way that they can be removed from the boards or replaced with ease. Hooks are used to hang displayed objects on the perforated or mesh boards. Attaching the objects to wires or string allows the visitor to take the object in their hands to examine the back or inside while preventing the loss of the item.

D. Mesh hangers or Velcro. These are used for displaying small objects.

Measuring instruments and rulers. These are used for analysis and data collection.

Miscellaneous tools. Colored string and tape can be used for grouping in layout and also to show relations between the displayed objects and posted data (see Figure 4-14).

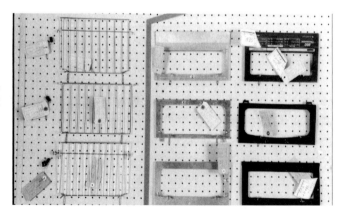

Figure 4-13
An example of display
on a perforated board

Figure 4-14:
An example of display
on a mesh board with
Tapes and tags used in
Tear-Down

Tags and color seals. Tags and seals are used to identify products and parts suppliers. Where tags are larger than the parts displayed, color-coded circular seals can be used for identification (see Figure 4-15).

Display Layout

Place a layout of the Tear-Down Room near the entrance. Assign knowl-

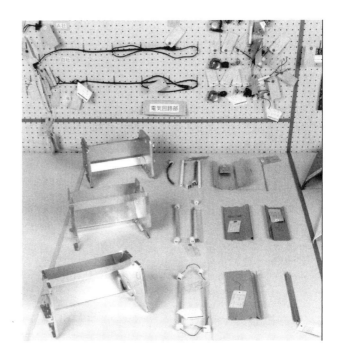

Figure 4-15

edgeable individuals to guide visitors through the display area and to respond to visitors' questions.

The movement of visitors is just like flowing water. Avoid making blind alleys.

Each aisle should be wide enough to allow people to pass behind those who are examining the display.

The display must be divided into blocks. For example, for an automobile, the display would include the engine and electronic equipment blocks. This will make it easier to understand the relationship between parts.

Use tapes and colored string to separate blocks as shown in Figure 4-15. This not only makes identification easier, but also helps visitors conceive of ideas relating to each block. Using distinct colors for different blocks also helps to separate the individual blocks.

Title each block to help people understand the elements of the display.

Announce the location of the display room, its entrance, and products on display. Direction signs should be placed to guide people to the display area.

Level of Disassembly of Products Displayed

The intent of Tear-Down is to disassemble and compare. Regardless of the extent of disassembly of the products displayed, parts identification must be maintained.

The relation among the assembly, subassembly, and each part must be readily visible. Ideally, the three levels should be displayed simultaneously. If that is impractical, parts lists and photographs should supplement the parts on display to help understand the issues. Available product and part drawings are also helpful

Parts Identification

Each part must always be identified with a tag or a circular colored seal. Information such as the part number, description, cost, weight, etc., posted on the tag or seal, will be helpful to visitors.

Suitable posters showing the relation between the colors of the tags are placed at the entrance and at suitable intervals in the display room. These posters help the visitors understand what they are viewing.

Information should be shown in graphs, rather than numbers. Purchased parts should also identify the suppliers.

To focus on differences between our product and competitors' products, products being compared are placed next to each other.

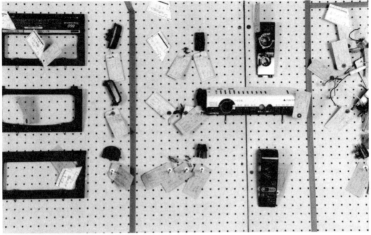

Figure 4-16 Display Divided Into Blocks.

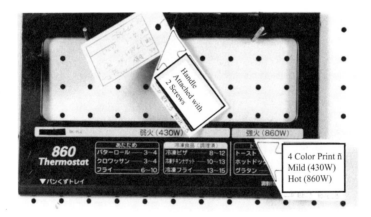

Figure 4-17 Stress What is Important

Appealing Points

Appeal to the visitor's sense of vision. Because the people concerned have analyzed the products displayed, the display should appeal to the visitors (see Figure 4-16). This will help visitors focus on those areas where ideas are being sought. Showing the points of attention clearly helps the visitors share views with the team.

Written information put up on the boards, including the results of analysis, must be linked with the product parts in order to guide the visitor's attention to the problem areas. Applying the same color marks, tape, or colored string to connect the products and related information is strongly suggested.

Highlighted product features and analytical information must be shown

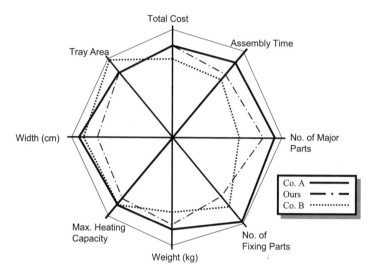

Figure 4-18 A Radar Chart Comparing Three Oven Toasters

in table or chart of comparison form to make the competitive features conspicuous. Samples of such charts are shown in Figures 4-17 and 4-18.

Types of Suggestion Sheets

Do not insist that the idea be presented on a standard, complicated suggestion form. This will discourage people from submitting ideas. Anything, including e-mail memos, can serve as a suggestion sheet. The objective is to encourage and collect as many suggestions as possible, not make it easier to process suggestion forms.

Have simple forms available which the visitors can complete while in the Tear-Down Room.

Anyone may write anything in the suggestion sheet, but sometimes what is written on the sheet is hard to understand. Team members should be available in the display room to discuss visitors' ideas and, if necessary, help them record the idea on the suggestion form. All parts on display should be properly labeled to help visitors not familiar with the product identify the parts affected by their ideas. Encourage the visitors to illustrate as much of their idea as possible so that their suggestion sheets are easy to understand.

Display information in a way that inspires ideas. Conversely, minimize information that would discourage ideas, such as past ideas that failed, unsubstantiated complaints, etc., unless such negative information is directly related to the areas needing improvement ideas.

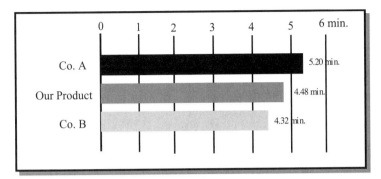

Figure 4-19 Graph Comparing Total Assembly Labor Hours for Oven Toasters

Displaying the Suggestion Sheets

Handling of the suggestion sheets is usually the responsibility of the team recorder, who passes them on to the next process team. One good idea is to design the forms as tags available in the display. Using this tag system, which was developed by General Motors, the suggestion sheet tags are titled in and attached to the product displayed.

Advantages

Duplication in suggestions is reduced.

Posted suggestions can inspire new suggestions, sometimes better suggestions than the original are created.

Disadvantages

Some people are content with the suggestions already turned in, and are reluctant to add or modify suggestions.

Reading other's suggestions sometimes intimidate other suggestors. Some visitors may think, "My idea can never be better than any of those submitted," and fail to express their ideas.

CLOSING POINTS

The following points are closing guidelines offered when conducting Static Tear-Down:

A. Create environments that attract many people. The Tear-Down display must be attractive in themes, focus, and appeal points, and offered in a relaxing environment..

B. Encourage visitors to invite associates.

C. Send invitations to all people concerned.

D. Appeal to the visitor's visual senses. Avoid displaying written technical reports. As mentioned previously, "Don't let them read, let them see." This can be improved to "Don't let them see, let them understand." Make the most of graphs and charts to achieve this purpose.

EVALUATING VA TEAR-DOWN RESULTS

The VA Tear-Down method compares our products with our competitors' products, focusing on differences relating to cost and function both for product components and for the entire product. The purpose of VA Tear-Down is to uncover competitor advantages, develop ideas to eliminate such advantages, and surge ahead of competition.

Even if the VA Tear-Down team successfully isolates the reasons for higher cost than the competition and takes corrective actions to correct the prob-

Date		Team:											
Function Description		**Comparative Evaluation**										TOTAL	RANK
		A	B	C	D	E	F	G	H	I	J		
A	Appearance		1	0	0	1	1	1	1	0	1	6	4
B	Projected Area	0		0	0	1	1	1	1	0	1	5	5
C	Switch Operations	1	1		1	1	1	1	1	1	1	9	1
D	Food Placement/Retrieve Ease	1	1	0		1	1	1	1	1	1	8	2
E	Control Panel Visibility	0	0	0	0		1	0	1	0	1	3	7
F	Caution Labels	0	0	0	0	0		0	0	0	0	0	10
G	Cleaning Ease	0	0	0	0	1	1		1	0	1	4	6
H	Crumb Removal	0	0	0	0	0	1	1		0	1	2	8
I	Cooking Area	1	1	0	0	1	1	1	1		1	7	3
J	Cord Length	0	0	0	0	0	1	0	0	0		1	9

Figure 5-1   FD Analysis of Toaster Oven Functions

lem, these steps do not assure that the competitive disadvantage has been resolved.

THE CUSTOMER DETERMINES VALUE

Whether our products are cost competitive or not can be determined with Cost Tear-Down. But whether a product is more valuable and, therefore, worth more to the customer cannot be determined unless the relationship between the product's cost and functions are resolved. Customers are not interested in products if they are less expensive but functionally inferior to competitor's products, or if they lack wanted features. Products can no longer prevail in today's market based on price alone. Conversely, a product that is disproportionately costly in relation to its functions will also turn away customers because the producer will have failed to achieve the balance of function and cost that attracts the major customer base. In other words, cost and functions must be properly balanced. The value of a product lies in this balance. Customers evaluate a product by its value. Value Analysis expresses this balance as;

$$\text{Value} = \frac{\text{Function}}{\text{Cost}}$$

Because it is the customer, not the producer, that determines value, the value of products must be seen through the eyes of the customer. It is, therefore, strategically important to determine not only how value is added, but also how much our product is different in value from those of the competitors.

Marketing can offer valuable information regarding customer preferences by conducting interviews, organizing focus groups, and surveying selected markets. Other, more analogical ways to rank functions and features by customer preferences are also used to determine and rank product features that add value, as perceived by customers. The Value Analysis equation (on page 22 and 33) is used to measure value. Determining cost (or price) is relatively easy to calculate. Quantifying functions is far more difficult. The goodness of functions depends largely on the needs, wants, individual preferences, timing, environment, and many other factors. Beer may be good and valuable in hot weather after strenuous activity, but may be less valuable in mid-winter.

There is no absolute way for producers to determine customer value perception. But VA Tear-Down poses the question and helps companies make wise decisions. It is recommended that a marketing or sales representative, familiar with the product being analyzed, join the team in conducting the analysis of value.

WEIGHTING EACH EVALUATION ELEMENT
Select Items Where Competitiveness Can Be Compared
There are many aspects to competitiveness, and many aspects of a product that have to be compared to determine its competitiveness. In addition to cost and functions, the following issues must be considered:
1. What do the customers want?
2. What is it about the product that attracts customers?
3. How much are customers willing to pay for the functions and features of a product?
4. What superior product characteristics will motivate customers to purchase a product?

The results of discussions and analysis will be a list of functions that the customer base wants or needs in the product. However, all the product functions and characteristics are not equal.

To add all the value characteristics to a product may make the product too expensive for the purchaser. Therefore, a priority ranking of those characteristics must be established, one that will offer the customer those high-value functions as a trade-off for those of lesser value.

Prioritize Items Where Competitive Advantages Can Be Compared
If the priority for each item can be assessed and quantified for comparison, functions can be objectively measured. A method for rating the priority for each function, and then expressing that priority numerically, is now introduced.

MEASUREMENT METHODS
FORCED DECISION (FD) METHOD
In this evaluation, items are compared in pairs as in a league tournament. Of the pair, the one item felt to have the greater value to the customer is given the score of 1 and the other is given the score of 0. Referring to the oven toaster example (see Figure 5-1) when Appearance and Projected Area are compared as to which is the greater customer value, Appearance is judged by the evaluators to be the more important value of the two and is scored 1. Projected Area, therefore, receives a score of 0.

	Function Description										J	K	W
1	Switch Operations	1.2									1.2	8.2	20.7
2	Food Placement/Retrieve Ease	1	1.2								1.2	6.8	17.2
3	Cooking Area		1	1.1							1.1	5.7	14.4
4	Appearance			1	1.1						1.1	5.2	13.1
5	Projected Area				1	1.3					1.3	4.7	11.9
6	Cleaning Ease					1	2.0				2.0	3.6	9.1
7	Control Panel Visibility						1	1.2			1.2	1.8	4.5
8	Crumb Removal							1	1.4		1.4	1.5	3.8
9	Cord Length								1	1.1	1.1	1.1	2.8
10	Caution Labels									1	1.0	1.0	2.5
	Total											39.6	100

Date   Team:

Figure 5-2   DARE Analysis of Toaster Oven Functions

Next, Appearance and Switch Operations are compared. In this comparison, Switch Operations is judged to have greater value than appearance. Now Appearance receives a 0 and Switch Operations is scored 1. When all the pairs have been compared, the total scores given to each of the items are summed in the Total column. Priority is given to each item according to its total points, as shown in the column titled Ranking.

Whether Switch Operations has a greater customer value than Appearance could be debated. When assessing the switch on a toaster oven, Switch Operations was judged to be more valuable to the customer. However, a paired comparison of a wall switch would find that Appearance has more value to the customer than Switch Operations.

The reason for the reverse value is that a wall switch is simple, where ease of operations is a transparent expectation. The wall switch is also a stand-alone item and part of the room decoration. The switch on the toaster oven can be more complex in turning the toaster on and off, as well as setting the heat range. Also, as an integral part of the toaster oven, the switch appearance is actually a small part of the overall appearance of the toaster oven.

Although appearance is valued higher than switch operations in making a buy decision, the customer will return the toaster oven if the operation of the switch is complex.

Judging customer value must consider the specific environment of the functions being evaluated, rather than generalize the customer value preferences.

DECISION ALTERNATIVE RATIO EVALUATION (DARE) SYSTEM

Using the FD (Forced Decision) method, the priority of each function to be compared is determined by anticipating which of the two in a pair of functions the customer would select. However, the importance of each function also needs to be determined. Simply because two items are close to each other in priority ranking does not indicate that they are almost equal in importance. DARE is the method to quantify the degree of importance.

It is difficult to determine how important a function is in relation to all other functions. But once the priorities for all items have been determined by the FD method, it is relatively easy to compare the importance of two items that are next to each other in priority. Beginning with the item highest in priority ranking, determine how much more important that item is than the item that is directly below in priority ranking. The item lower in priority is given the factor 1, and the higher priority item a number higher than 1.

Returning to the oven toaster example, as shown in Figure 5-2, it has been determined that Switch Operations is 1.2 times more important than Food Placement/Retrieve. This value is entered in column J. The process is repeated for all the prioritized functions.

The values in column J are subjective. They are the result of the team discussing the comparative values of the functions compared to the lower valued function. As an example, the FD method determined that Cord Length was more important than Caution Labels, but by how much? Starting with the lowest value, the team decided Cord Length was 1.1 times more important than Caution Labels. Next the team discussed the relative importance or degree of separation between Crumb Removal over Cord Length. The team decided Crumb Removal was 1.4 times more important than Cord Length. The Tear-Down Team continued with their assessment.

Once all pairs have been compared, the degrees of importance for all the function items are converted into values calculated on the basis of 1.0, the lowest valued attribute. The ratio of importance between paired functions is calculated and entered in column K. For example,

[Caution Labels/Cord Length] X [Crumb Removal/Caution Labels] =
1.1 x 1.4 = 1.5 (1.54 rounded)

Multiplying [(1.1 X 1.4) X 1.2] = 1.8, the K value of Control Panel Visibility over the preceding two functions. Continuing with the calculations in column K, the degree of importance for Ease in Cleaning is 3.6 and Appearance is 5.2 when compared with all the other items.

It is common practice to convert the absolute values of Degree of Importance from column K to percent. In this case, 100 percent is assigned in column W to the absolute value total of 39.6 from column K. By converting the data to percent, the values are normalized and will relate to any product example using DARE. The values calculated this way are called Degree of Importance Factors; these can then be applied to many areas including cost distribution.

PAIRED COMPARISON

The Paired Comparison shown in Figure 5-3 combines the FD and DARE processes in one continuous step. Paired Comparison is a simpler alternate method to the FD and DARE processes. Either FD and DARE together or Paired Comparison can be used at this point in the evaluation, depending on the evaluating team's preference.

Referring to Figure 5-3, the ten function descriptions are alphabetically listed. As in the FD method, the functions are evaluated in pairs to determine the valued preference of the pair. For example, comparing Appearance to Projected Area, the team selected A (Appearance). The team was than asked, "By what degree of importance is function A better than function B?" The team has three choices (see Choice Factors). If the team found it difficult to choose between A and B in determining which function was better, the team would select choice factor 1. If, after some debate, the team agreed Appearance was better, the choice factor would be 2. If the team selected Appearance after a short discussion, if any, the choice factor would be 3.

In this example, the team selected choice factor 3, which appears in the first square as A3.

Comparing function A to function C (Switch Operations) C was chosen with a choice factor of 2 (see cell C2). After Appearance has been evaluated against the other functions, Projected Area (B) is compared in a like manner. The

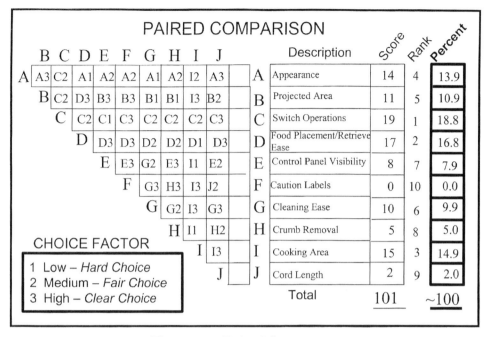

	B	C	D	E	F	G	H	I	J		Description	Score	Rank	Percent
A	A3	C2	A1	A2	A2	A1	A2	I2	A3	A	Appearance	14	4	13.9
B		C2	D3	B3	B3	B1	B1	I3	B2	B	Projected Area	11	5	10.9
C			C2	C1	C3	C2	C2	C2	C3	C	Switch Operations	19	1	18.8
D				D3	D3	D2	D2	D1	D3	D	Food Placement/Retrieve Ease	17	2	16.8
E					E3	G2	E3	I1	E2	E	Control Panel Visibility	8	7	7.9
F						G3	H3	I3	J2	F	Caution Labels	0	10	0.0
G							G2	I3	G3	G	Cleaning Ease	10	6	9.9
H								I1	H2	H	Crumb Removal	5	8	5.0
I									I3	I	Cooking Area	15	3	14.9
J										J	Cord Length	2	9	2.0
											Total	101		~100

CHOICE FACTOR

1 Low – *Hard Choice*
2 Medium – *Fair Choice*
3 High – *Clear Choice*

Figure 5-3   Paired Comparison

paired comparison scores are totaled and normalized in the Percent column. The Rank column places the functions in descending order of importance, based on the function's score. The Percent column in Figure 5-3 corresponds to column W in the DARE analysis (Figure 5-2). Differences in values are the result of differences in approaches. However, the important relationships are the same.

VA Tear-Down members are encouraged to select and modify the FD, DARE, and Paired Comparison processes to suit the purpose and products selected for evaluation. A modification for the Paired Comparison method is to expand the choice factors from three degrees of importance to five. This will result in a greater degree of separation in the relative importance of the functions. It would not, however, change the function ranking.

A brief reference is made to another method of evaluation with the emphasis on customer affordability. This method evaluates the value of each function from the viewpoint of how much money customers are willing to pay for each function.

In the case of the toaster ovens, the value of the function of toasting bread

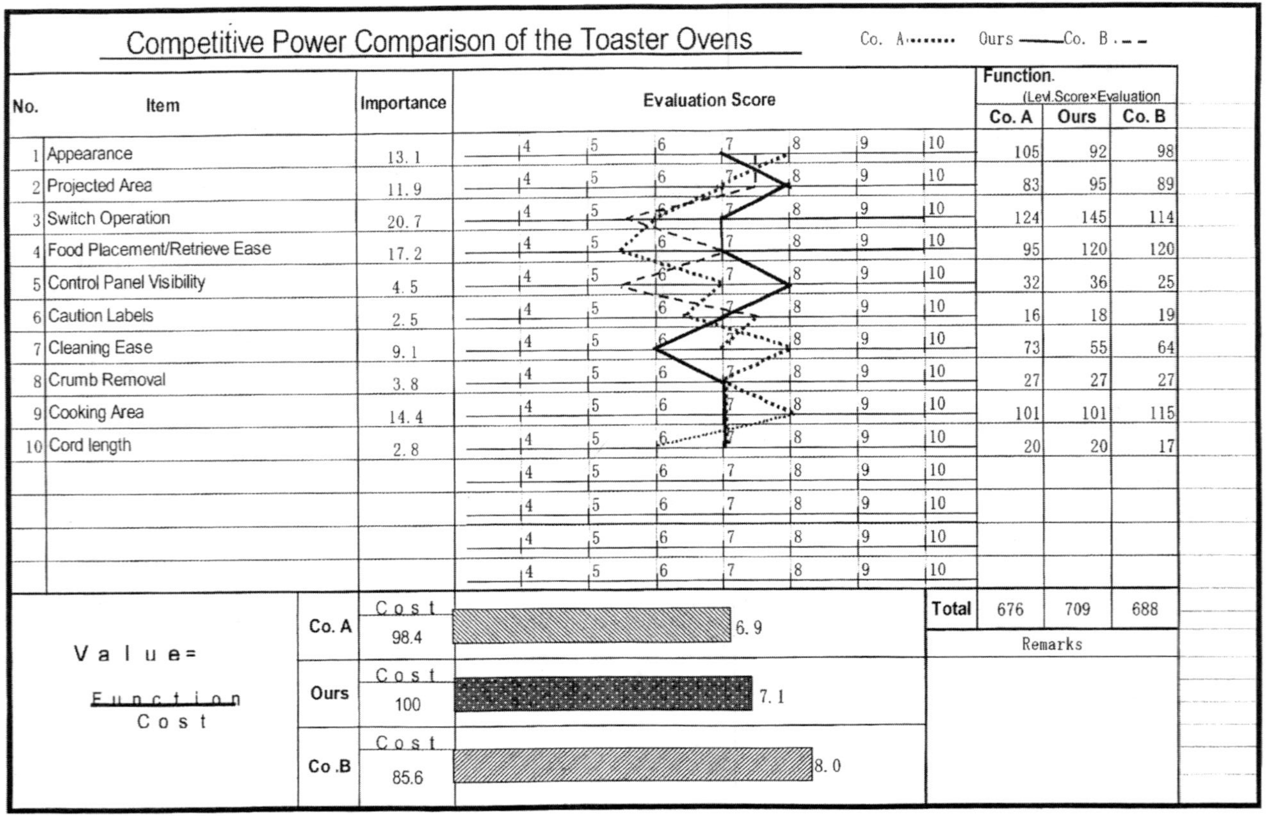

Competitive Power Comparison of the Toaster Ovens

Co. A ·········   Ours ———   Co. B - - -

No.	Item	Importance	Evaluation Score	Function (Lev.Score×Evaluation) Co. A	Ours	Co. B
1	Appearance	13.1		105	92	98
2	Projected Area	11.9		83	95	89
3	Switch Operation	20.7		124	145	114
4	Food Placement/Retrieve Ease	17.2		95	120	120
5	Control Panel Visibility	4.5		32	36	25
6	Caution Labels	2.5		16	18	19
7	Cleaning Ease	9.1		73	55	64
8	Crumb Removal	3.8		27	27	27
9	Cooking Area	14.4		101	101	115
10	Cord length	2.8		20	20	17
			Total	676	709	688

Remarks

$$Value = \frac{Function}{Cost}$$

Co. A	Cost 98.4	6.9
Ours	Cost 100	7.1
Co. B	Cost 85.6	8.0

Figure 5-4   Toaster Oven Competitiveness Comparison Chart

is equal to the price of a toaster without the oven. We assume the lowest price of such a toaster on the market (market value standard) to be about $16.00. What then is the value of the oven function? Evaluation is made for such functions as visibility, capacity, and range of applications. Because these are existing products, the actual market retail price may be used for the market value. Customer focus groups and surveys may help to determine market value.

In a study involving a headlamp wiper on a recreational vehicle, the result of a survey was almost equally divided between those who favored the wiper and were willing to pay a relatively high price, and those who saw no need for it and rated its value zero. Nevertheless, surveys are a useful means of collecting information because they quantify user feeling. Proper and well-designed questions will serve the objectives of the VA Tear-Down process.

COMPETITIVENESS ANALYSIS

The preceding section stressed the importance of each function offered by a product. It is equally important to determine how well these functions perform when compared to the competition's products.

To make this comparison, team members evaluate each function on a scale from 0 (worst) to 10 (best), with 5 representing the standard, and then calculate the average values. The evaluation must be made as objectively as possible from the standpoint of the products' consumers. Exclude people who were involved in the creation of the product. Product engineers have a bias favoring their creation and will protect their design decisions. What are needed are representatives of actual purchasers and users; these representatives can judge the value of the functions offered, their relative goodness, and their willingness to buy the product.

COMPETITIVENESS COMPARISON CHART

Although the values for individual function items are known, they are not equal in importance, and cannot be simply summed up. The rating by each evaluator is multiplied by the degree of importance (see Column W in Figure 5-2) to obtain the function point for that item. Then the ratings by all the evaluators are summed up as an average score for that function.

The numerical score does not indicate an absolute value. When compared to the competitor's score for the same function, however, the score indicates the relative ranking of the products being compared.

Referring to the toaster oven example in Figure 5-4 the rating given to Appearance for Competitor A's toaster is 8 (the average of the evaluator's score) and the degree of importance factor is 13.1 (see Figure 5-2, Column W). Therefore, its function point is 105 (8x13.1). Similarly, the function points are calculated for all function items. The total of the function points is 676 for Competitor A's toaster oven, 709 for our toaster oven, and 688 for Competitor B's toaster oven.

In addition to the total function points, it is also important to compare the points for each function. Comparison by function makes it possible to determine the functions for which our product lags the competitors, that function's importance, and the priority of the effort to catch up with or exceed the competitors. All people concerned can easily share the analysis and the status as well as what has to be done.

The chart shown in Figure 5-4 also helps develop sales strategies. For example, our product does not necessarily have to prevail over all others in all of the function categories. By developing superiority in functions highly valued by customers, we can advertise and promote these functions to gain market share.

The information displayed in Figure 5-3, Figure 5-4 and similar charts, added to the parts display in Static Tear-Down, help assure beneficial results from the use of VA Tear-Down.

VALUE EVALUATION

The function evaluation is one part of the value equation. Even if our product scored superior to its competitors, the issue remains: Is the customer willing to pay for those functions?

As noted above, the value equation (value = function/cost) is used in Value Analysis to measure value. The cost part of the equation has been determined previously in the Cost Tear-Down. Competitor A's cost index is 98.4 and B's index 95.6, with our cost set at 100. The value (function relative to cost) can be calculated with the total of the function points divided by each of these indices. The resultant values are used to assess the value of our product in relation to competitors' products.

The values of the three toasters calculated by means of this equation are shown in Figure 5-5. The same figure is shown at the bottom of Figure 5-4. The value of competitor B's toaster oven is calculated by dividing the total of func-

	Cost %	Function Point	Values	
Product A	98.4	676		6.9
Ours	100	709		7.1
Product B	85.6	688		8.0

Figure 5-5 Toaster Oven Price Comparison 1
(Calculated based on theoretical cost)

	Market Price (a)	Function Point (b)	Value (b/a X 100)s	
Product A	$36.80	676		18.4
Ours	$40.80	709		17.4
Product B	$38.40	688		17.9

Figure 5-6 Toaster Oven Price Comparison 2
(Calculated basis on market price)

tion points (688) by its cost index (85.6) to get a value of 8.0. This indicates that it is 12.7 percent better in value than our product, which has a value of 7.1. The value of competitor A's toaster oven, determined in the same way, is 6.9, which indicates it is slightly lower in cost but its low function point detracts from the advantage of the lower cost. Our toaster oven is in the middle in terms of value.

The same relationship can be applied by substituting Price for Cost. This is a more valid value approach because the customer cares little about the cost of the product. The customer's concern is the price of the product, or what the customer must pay for the functions offered.

In Figure 5-6, the function points for our product, 709, are divided by $40.80, which is the actual retail price rather than our suggested retail price of $56.00. The result is its market value index 17.4. Likewise, Competitor A's toaster oven is sold at $36.80, although its suggested price is $44.00. Therefore, its market value index is 18.4 (676÷36.8) which is 106% compared to our toaster oven (18.4÷17.4). Competitor B's toaster oven, which is retailed at $38.40 rather than B's suggested price of $52.00, has the market value index of 17.9 (688÷38.4), or 103% of ours.

The evaluation process described above is more representative of the customer's sense of value than only comparing costs, or the suggested retail prices. In today's market, most consumer hard products are sold at a discount against the manufacturers' suggested retail prices. This sales strategy is designed to increase the customer's sense of value because with identical function, a discount usually increases the customer's value perception. Customers inherently sense that the discount has caused the relative value to increase. Of the three toasters, customers are most likely to find competitor A the best value on the market in overall evaluation.

A word of caution: People do not view discounts and functions in the same way. They may feel that the manufacturer is conducting a sales offensive or trying to sell dated products. Customers may feel that the manufacturer is trying to reduce a product model inventory to make room for an upgraded model. Should people sense that the discount is being offered as a prelude to announcing a newer, better product, then the discount strategy may fail.

Another aspect of customers' psychology is that they often have an upper limit of prices that determines their buy, no-buy decision. Therefore, a product rated as high value may only appeal to a limited market segment because the price point is too high for many people.

OTHER MEASURES OF COMPETITIVENESS WITH VA TEAR-DOWN

The VA Tear-Down processes already described in this book rely on disassembling products, then determining and analyzing their differences. This concept can be applied to many aspects. Two of them are described in this chapter. One (the Mona Lisa method) applies VA Tear-Down to function development for improving competitiveness; the other (the Unit Cost method) provides objective measures developed from comparison and analysis.

THE MONA LISA METHOD

The term *Mona Lisa* is applied to this process as an expression of its beauty and method of deduction in determining customer preferences.

The VA Tear-Down processes described in previous chapters focus on reducing cost and improving functions in order to increase value. The common denominator of any Tear-Down process is comparison. Applying VA Tear-Down to the complete product will help determine and understand the customer's motivation for selecting a product by identifying the relative value of product functions. The Mona Lisa method introduced here is an application of VA Tear-Down; it develops greater value from function improvement by concentrating on the customer's most-valued functions.

The Mona Lisa method described below has been adapted from value analysis and marketing procedures as a function-developing VA Tear-Down approach. This approach uses comparative analysis to find:

What functions and quality levels are most highly valued by customers?

Why are these functions valued?

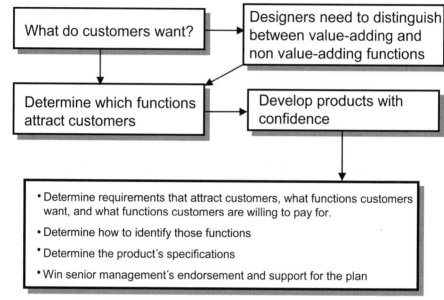

Figure 6-1 What Is Mona Lisa?

ESSENTIALS OF THE MONA LISA METHOD

Many manufacturers have made significant investments in market research, attempting to determine what customers need and want, as well as how much they are willing to pay. In today's global market this information is essential for developing a market strategy that covers market share growth, sales, programs, profit, and payback on investment in product development. Whether a product sells or not is finally decided on the product's appeal to individual customer's sense of value and customer satisfaction with the product after the sale is made.

Each manufacturer plans new products on the basis of market and competitive intelligence, plus future prospects. Unique features provide a way for a manufacturer to separate its products from the competition. Although there is a need for uniqueness, the result is not always encouraging; sometimes, the manufacturer puts too much emphasis on incorporating unique features or functions, that may not add value from the customer's point of view.

Functions classified as required are often placed in the need category, indicating that the function must be incorporated in the product. However, sales often classify all competitive product functions as needed. The result is a range of functions from basic functions that must be satisfied to enhancement functions, such as improving comfort, that are much lower in priority. When a new competitor appears, new functions in the product are often mistaken as a need by the competition, thus blurring customers' real requirements. Such errors in judgment are costly. They can be avoided by comparing and analyzing competing or similar products that are on the market, then selecting the functions and features that most satisfy customers (see Figure 6-1).

Positioning Mona Lisa

What the customer likes is not the only consideration in developing product specifications. In addition, capabilities and resources in development, production, and marketing need to be considered. Four major factors are involved in deciding on specifications, as shown in Figure 6-2.

Customer Valued Functions and Quality refer to identifying those factors with attributes that attract customers. Many of those attributes are value perceptions. Attributes such as style, ease of operations, after-market service, and price are examples, among others, of value attributes stimulated by effective marketing and advertising.

Market Trends is another influencing "buy" factor. If trends such as style changes are dynamic, the customer may delay purchasing the product for a later model. Marketing strategy is a major influence for determining if dynamic, or style changes are more profitable.

New Materials, Processes, Design and Other Trends are factors that influence ts the decision either to enter the market with a new product, or to offer a less dramatic model of the same product. That choice is governed by the rate of technology advancement.

Business Planning, Sales Strategy, Cost, and Investment are all factors that control the product development. As such, the major issue controlling all decisions affecting the product is; "Will the proposal contribute to the growth and profitabilltyprofitability of the company?" It is critical, therefore, critical that all product plans be presented to management in a form of a business case.

As shown in Figure 6-2, once the decisions mentioned above have been resolved concerning the product, it is up to the Engineering Staff to convert ideas to reality.

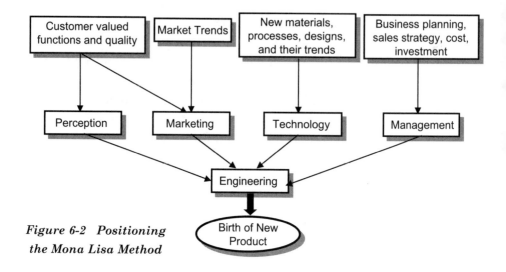

*Figure 6-2 Positioning
the Mona Lisa Method*

BASIC STEPS

The Mona Lisa method, used with the VA Tear-Down process, is performed in steps that are similar to other Tear-Down procedures. These steps are shown in Figure 6-3.

A. Plan

As with other Tear-Down procedures, the plan must be established on the basis of 5W1H, focusing on:

1. What specific product improvements are needed?
2. What should the scope of the investigation cover?
3. What should be done by when, to develop the product as scheduled. The schedule should be determined by working back from the end point.

B. Select Products

Product selection is a major point in the Mona Lisa process. In Mona Lisa, as with any other Tear-Down procedures, the results are largely dependent on the selection of products to be compared, and how advantages in the products selected translate into suggestions for improvement of our own products.

1. Make sure that one of the competing products selected for evaluation is the market leader.

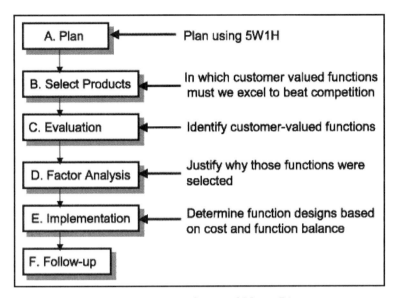

Figure 6-3 Basic Steps of Mona Lisa

2. Include innovative competing products.

3. One competing unit is usually sufficient for evaluation of many types of small products, including home electric appliances. For larger products such as automobiles, two units each are recommended, one for disassembling and display, and the other for road handling characteristics. A rented car can be used for the latter purpose.

C. Evaluation

With an understanding of the functions and features that attract customers, and their priority, the process to determine how the competing products incorporate these functions can proceed. Evaluators now assume the role of the customer in operating the product and maintaining it.

1. Personnel directly involved in developing the product should not participate because they have a protective bias towards their creation. Men and women who represent the actual users of the product and who can perform objective evaluation must be chosen.

2. The number of evaluators should be a minimum of 15-to-20 participants to achieve a statistically valid survey. The number of participants depends on the product being sampled, but at least 15 or more people are desirable.

Mona Lisa Evaluation Form

Name of project:_____ Date:_____

Welcome to Mona Lisa. We are seeking your objective views to assist us in developing a new product. Please assist us by filling-out this form in accordance with the instructions below.

Evaluation Procedures:

* Use one set of evaluation forms for each product

* Evaluate the items shown below by selecting an evaluation that best expresses your feelings.

10. Excellent	4. poor
8. Good	2. Bad
6. Fair	0. Unacceptable

Please translate your perception into any number from 0 to 10, including odd numbers. Please give your reasons for any item rated 7 or more, or rated 4 or less.

Product Evaluated _____ Manufactured _____ Model_____

Your name _____ Dept._____ Phone No. _____

Oven Toaster

APPEARANCE	Rating			Rating
1 Overall appearance		4	Warning labels readability	
2 Door Operations		5	Overall Size	
3 Inside visibility		6		
SWITCH OPERATIONS				
1 Timer ease of operations		4	Cooking speed	
2 Switch controls		5	Cooking complete signal	
3 Power plug insert ease		6		
Maintenance				
1 Clean window ease		4	Inside cleaning ease	
2 Crumb Removal		5	Cooking complete signal	
3 Power plug insert ease		6		
1		4		
2		5		
3		6		

You may be contacted and requested to meet with the team to share your perceptions with the team. Please include your name and phone number where you can be reached.

Figure 6-4: Example of Mona Lisa Evaluation Form

3. In the process of evaluation, it is important to note what is good, what is bad, why, how good or bad, and what factors influenced these decisions.

The following is the essence of the evaluation processes.

a. Static evaluation: Subjective evaluation

b. Dynamic evaluation: Evaluation of ease in operation and serviceability

c. Evaluation of goodness is scored using a relative ranking scale from 1 to 10.

        10: Excellent

        8:   Good

        6:   Fair

        4:   Poor

        2:   Bad

        0:   Very bad

The items that have scored 9 or more on the average are called "Mona Lisa items," those scoring 7 or more are attractive items, and those scoring 4 or less are screened out of the evaluation processes.

Examples of Mona Lisa evaluation forms are shown in Figure 6-4. These forms describe the information collected and the scoring process. Each company should develop its own form and questions to best represent its product, market, and business strategy.

D.   FACTOR ANALYSIS

Having learned the evaluator's perception of how good or bad certain functions of a product are performed, it is important to determine why the evaluators scored the functions as they did. Suppose the shaving head on a competitor's electric razor received a high score. Before any product improvements can be suggested the reason for the high rating needs to be determined. Was the high score the result of ease of use, cutting performance, size of the cutters, cleaning ease, or unique design? To replicate the user's experience, the Tear-Down team members need to use and maintain the product themselves before recommending value-adding changes.

The next step in the process is to analyze peripheral factors that influence competitor's high scores and also affect the suggestions and proposals that improve the product.

Questionnaire Analysis

First, determine who evaluated the item or who answered our question-naire. For example, the fact that an item was well accepted by women but not by men would be an important factor in the analysis. Although all opinions are valuable, for best results people representing the product's market segment should participate in the survey.

1. Functions scoring high are sorted from those of lower scores. High-scored functions will receive the team's attention in determining fur ther actions. Those functions rated on the low end are set aside because investing in functions that have little customer appeal would be a poor business decision.

2. With the scores tallied, we should compare them in each of the products evaluated to better understand the functions that appeal most to the customers. To better appreciate the customer appeal, the product should be examined in its assembled and partially disassem-bled condition.

3. Those design characteristics that resulted in functions having high customer-appeal scores are then analyzed to determine the cost incurred by these functions.

E.   Implementation

The objective of Mona Lisa is to create the best world-class product by dis-covering the functions most appealing to customers and then developing the best in design and layout. These goals are achieved through value analysis of present products, both ours and competitors'. However, the method would serve no useful purpose if the best product in the world should be the highest cost in the world. The result would be a very narrow niche market segment consisting of customers who are willing to pay for all the best functions. To avoid this situation, some cautions must be taken in proposal selection and implementation.

Items to be Selected

1. Functions in the Mona Lisa category are initially accepted without regard to their incurred cost, but cost consideration must follow.

2. Sort the attractive functions in the order of priority with their esti-

mated incurred cost. Preliminary selection is made at this time within the framework of the budget.

3. Select functions as trade-off candidates for existing functions.

4. Include in the selection factors the affordable cost, including redesign, to achieve the selected functions.

Action Plan Factors to Consider

1. Study the lowest cost way to implement the selected functions.

2. In this attempt, make the best use of Value Analysis and VA Tear-Down techniques.

3. To broaden the scope of investigation and the number of ideas, include additional competitive products to those previously selected and analyzed in the VA Tear-Down process.

F. Follow-Up

Mona Lisa procedures are likely to result in a large number of ideas that may incur unacceptable costs. Implementation plans must be based on a careful balance of popular functions and their incurred cost. The VA Tear-Down team must, therefore, focus their full attention on ways to achieve the goals of adding the maximum value to the product with a minimum of cost.

1. Decide who will be included in the development of selected functions while minimizing the cost.

2. Provide the human resources needed to follow through with the plan. Make arrangements for back-up if necessary.

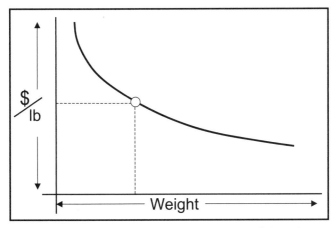

Figure 6-5 Cost Per Unit of Weight VS Weight

3. Check the progress of the plan against its schedule. Resolve any problems that may arise with any support needed.

4. Determine if the cost targets are on track, or if any additional actions are required to achieve the targets. A particular function may exceed its cost target, but other functions may under-run their targets. The net effect of incorporating all function improvements must meet the cost target.

5. Finally, check if the plan has been successfully completed.

THE UNIT-COST METHOD (PARAMETRICS)

THE UNIT-COST METHOD CONCEPT

Chapter 5 discussed procedures for measuring values. However, the result of the evaluators may vary from person to person impacting the reliability of the data, because many functions cannot be evaluated quantitatively. Functions such as aesthetics, customer appeal, and others are qualitative; they are scored on the basis of relative comparison.

How can we measure if a function is performing good or bad, or how good or how bad? In a search for such a measure, a simplified method of evaluation called the unit-cost method can be used.

In this method, a unit is calculated using elements (in this case, cost and weight) that can be converted into numerical values. Evaluation is made on the basis of the value of that unit. In this example, the unit is the cost per unit of weight, that is, $/lb. Depending on the product and market, other objective units can be selected such as kilowatts per hour, drilling depth per hour, and cost per capacity. You can, in fact, select combinations of related measurable elements most appropriate for a given purpose. For example, a drilling machine that drills holes for placing explosive charges used to remove earth for building highways received accolades at the annual equipment show for the innovative design appearance. The drilling machine could turn in a small radius; it had an air-conditioned cab, digital instrument display, and many other attractive features. However, the product failed to meet sales objectives because the most valued function was measured in terms of cost per foot of hole drilled. This highly-valued performance measurement was not equal to that of its less attractive competitors.

In the discussion below, everything is converted into the relation between weight and cost. The result will be used as a measure to determine advantage or disadvantage. It is recommended that readers create measures to suit their own purposes.

OBJECTIVES

1. Identify abnormal values.
2. Coordinate the levels of prices or costs.
3. Weigh the possibility for improvement against set targets.

SCOPE OF APPLICATION

Anything that has more than one element can be compared (e.g., cost and weight).

Outline of the Unit-Cost Method

The equation in this example is:

$$\text{Cost} = (\$/\text{lb}) \times \text{weight}$$

The relation between the weight and the cost per unit of weight is shown in the graph of Figure 6-5. The major elements that affect either cost or weight ($/lb) are listed in Figure 6-6.

		Advantage	Disadvantage
1.	Weight	Heavy ——	Light
2.	Processing Steps	Low ——	High
3.	Number of Parts	Few ——	Many
4.	Number of Processes	Few ——	Many
5.	Surface Treatment	Little ——	Much
6.	Material Grade	low ——	High
7.	Logistic Line	Short ——	Long

Reducing weight causes $/lb to increase. However our objective is NOT to reduce $/lb, but to reduce cost.

Figure 6-6   Major Elements That Affect Cost Per Unit of Weight

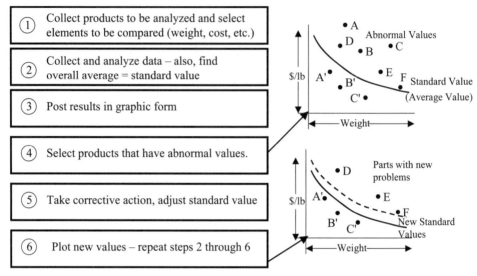

Figure 6-7 Analysis of Unit Weight / Cost Method

BASIC ANALYSIS PROCEDURE

The basic procedure of the unit-cost method is shown in Figure 6-7.

SPECIFIC ACTIONS

Analysis of Assembly

The first step in the analysis process is to collect the sample products in one area and identify the parts on the products to be analyzed. The complete assembly is analyzed first. The value of $/lb is calculated and Steps (2) through (5) in Figure 6-7 are followed. The data points shown on the graph are the result of comparing similar assemblies and components of the sample products that perform the same function. The standard value curve is configured by determining the least square fit of the data points, excluding the anomalies (Point A and point C). Even if some improvement has been achieved for the entire assembly, following step 5, there often remain some problems in the component parts. Therefore, the analysis of the assembly's components parts follow.

Analysis of Component Parts

The product or assembly is disassembled into component parts. Similar competitor parts performing the same function (e.g., castings, machined parts, and plastics parts) are analyzed on the same graph to find potential problems. As with the assemblies, steps (2) through (5) in Figure 6-7 are performed.

Analysis With Cost Elements

1. Processing cost, material cost, and other cost elements are selected from the component cost breakdown. They are divided by the pertinent weights, and the results are then plotted on a graph. The points on the graph in relationship to the standard value curve will indicate any problems.

2. When analyzing cost, special attention is paid to whether excessive processing is performed. The checklist points to consider are:

    a. Is the number of processes reasonable?

    b. Are too many parts used, causing excessive labor-hours?

    c. Is excessive machining applied (ranges, accuracy)?

    d. Is any other cost-effective method of processing available?

    e. Is unnecessary operation performed (tool changeover, inspection, machining applied to areas where it is not necessary)?

    f. Are unnecessarily high-grade machines or facilities used?

    g. Is inter-process transfer affected (dead time, transfer, stuck in process)?

    h. Is there any excess in logistics or packing (distance, excessive pack ing, stock cost)?

    i. Is the labor cost of purchased materials too high

Checklist Points in Material Cost

    1. Is the material of unnecessarily high grade?

    2. Is scrap rate within a reasonable range?

    3. Can scrap material be used productively?

    4. Is the level of material cost reasonable?

VA TEAR-DOWN SUMMARY

CREATE A MODEL CASE FOR SUCCESS

Starting a VA Tear-Down program with very high expectations can cause the project to fail. Start from a relatively simple or small project; then, as successful experience is achieved, gradually work up to larger, more difficult ones.

One danger in introducing VA Tear-Down in a company is the perception that the process can be used freely, in any situation, without the need for qualified people. To ensure success, however, the people assigned need to be capable of following through with the VA Tear-Down procedures. Before launching a large VA Tear-Down effort, it is strongly advised to start with a well-planned pilot program. This will serve to test the process, identify realistic expectations, and make any process and procedural adjustments prior to the official start of the effort.

ESTABLISH A LEVEL OF COMPETENCY

Like any other specialized activity, VA Tear-Down requires a core team of trained personnel dedicated to the proper and successful application of this methodology. Part-time, unskilled personnel will not achieve professional results.

ORGANIZATION

The most effective way to support the VA Tear-Down activity and expand it into other company units is through a corporate support unit. In Japan the unit is called the Promotion Center. This center of support coordinates the VA Tear-Down activities across the divisions and units of a company, and provides resources and investments to implement improvement activities. The promotion center must have a clear status in the company; it should function to mobilize and lead the entire company in Tear-Down activities. A company-wide, authorized organization must be established on its methodology merits, rather than on a personality. This arrangement would ensure that any change in personnel would not affect the continuity of the activities. Once such an organization is established, VA Tear-Down activities can be incorporated as an operating culture rather than a special event.

UNDERSTANDING BY THE TOP MANAGEMENT

Most company initiatives that survive are introduced and supported by senior management. Initiatives such as Quality Circles, Total Quality Management (TQM) and Six Sigma, all had their start through the endorsement of the company's senior officer. To succeed, VA Tear-Down requires the same level of support. Few initiatives enjoy longevity if, after starting on a unit level, it fails to attract top management support. Managers are attracted to activities that help them achieve their business objectives.

Although VA Tear-Down is designed to attract customers in selected markets, it is important also to address the concerns of internal customers, e.g., top management. An accepted invitation to top managers and their staffs to visit the tear-down display will allow the managers to use all of their senses in understanding the process and judging the results. A good VA Tear-Down manager will time the invitation to coincide with the top manager's business visit to the company. The VA Manager, supported by the unit manager, should have a carefully planned presentation showing the results of the study and how the process will support the corporate business objectives.

Three years after the co-author Yoshihiko Sato started the VA Tear-Down activities in an Isuzu auto division, he displayed the advantages of parts commonality that resulted from a Tear-Down project. The CEO of Isuzu Motors and other executives attended the display. The event won both the endorsement of top management and the introduction of VA Tear-Down across the Isuzu Corporation.

FACILITIES

The VA Tear-Down room should be designated as the center for improvement for the entire company. As previously discussed, the room should be permanent; it must be equipped with tools and supplies easy to use by employees, under supervision of a VA Tear-Down staff member. The location for the room should be convenient for anyone who wants to visit and it should have an inviting atmosphere.

STANDARDIZATION OF OPERATIONS

Once VA Tear-Down is integrated into the flow of daily operations, it will invite projects. The results will become visible, which in turn will make it easier to obtain necessary resources. As VA Tear-Down is promoted and people recognize its effects in improving operation effectiveness, saving cost, and enhancing product functions, support for the activities will increase. Without adding value, interest in VA Tear-Down will diminish. As a service support activity, its viability and survivability is dependent on the support of the people who benefit by the results of the methodology.

ESTABLISHING THE PLAN AND SETTING THE TARGETS

VA Tear-Down must become an active part of the business planning process, rather than be dependent on seeking unplanned targets of opportunity. Only by proving its worth in contributing to the success of the corporate and company business objectives will VA Tear-Down achieve the recognition and status it deserves.

VALUE ANALYSIS

The principles of Value Analysis have been integrated in all the VA Tear-Down processes described throughout this book. Value Analysis is a methodology; as such it is transparent, imbedded in VA Tear-Down, rather then identified as a process "step." In every VA Tear-Down project, Value Analysis is the methodology used in the steps of comparing and analyzing.

The foundation of Value Analysis rests on the identification and analysis of functions. Every action we take and decision we make is based on satisfying a function Using Value Analysis methods we can articulate the function we are addressing. Understanding, identifying, and prioritizing customer sensitive functions and therefore, their perception of value, is one of the keys to business success.

Refer to Chapter 2, and re-read the description and application of Value Analysis. Use this VA methodology as you follow the steps of the VA Tear-Down process.

APPENDIX WORKSHEETS

A variety of worksheets are included. They are intended as guides for preparing analysis forms that best represent the products and markets of the reader's company, rather than as templates that must be followed. Some of the examples mentioned specifically in this book depart from these forms because they have been adapted to fit the product and expected results.

An ideal worksheet is one that is designed to give meaningful and persuasive conclusions to the subject of the VA Tear-Down study. The reader is encouraged to study the conclusions that would represent a successful VA Tear-Down project, then determine not only what data must be collected in what form, but also how to interpret the resultant data.

There is a temptation to collect a massive amount of data on multi-page forms. This temptation should be avoided. Complexity leads to confusion. Simplicity adds credibility to the recommended value-adding proposals based on the data collected.

					Registration /Log in		
					Log No.		
TEAR DOWN PROPOSAL FORM					log date		
Name of Proposing Company			Date of proposal		Project	Exhibit Corner No.	
Name of Unit			proposer				
Proposal Titlle			Product from which idea is derived				
Name of the person making this proposal			Applicable products				
Delivery item no.							

Please detail, WHAT YOU ARE RECOMMENDING

Judgement			
Adopt/Non-adopt □ Adopt as proposed			
□ Adopt based on the comments in the right block.			
□ No good /not adoptable			
□ For future consideration (idea bank entry)			
□ Hold for future consideration			
Other Recommended Actions			
Dt/Mo/Yr of Review			

Producer's comments	Cost evaluation data		
		Name of evaluator:	

The proposer is required to fill in the framed portion.

A-1

IDEA MEMO			
Proposer		Log No.	
Part no.		Proposal date	
Part name			
Improvement Proposal	Proposal title:		

Describe your proposal in more detail:

Adopt/Non-adopt		Cost Evaluation	
(Adopt. condition/Non-adopt Reason)			
		Data transfer no. :	

A-2

| Dynamic Tear Down Man-hours Analysis Comparsion | Part | Cover Assy / Part Name | Re-port | Date | Section | Reporter |

Competitor — Sketch of Assembly Process

Our product — Sketch of Assembly Process

Competitor — Sketch of Assembly Process

Total Assembly Time

Process

REMARKS

Total Assembly Time · Total No. Fixings · Weight

Structural Elements & Related Assembly Factors

BOLT				SCREW			TAP SCREW			NUT				WASHER				CLIP				BAND			PIN		OTHER		
Non Washer	Flange	Single Sems	Double Sems	Non Washer	Single Sems	Double Sems	Non Washer	Single Sems	Double Sims	Flange	Single Sems	Resin	Spire	Plain	Spring	Lock	Welding	Anchor	Hose	Wire	Plate Band	Rubber	Resin	Metal	Split pin	Portable Jack?	Sealer	Bond	Tape

Co. A / Ours / Co. B

Comparison of Work Methods (by Assembly Factors)

Overall evaluation · Working posture · Working space · Working position · Identification · Torque control · Tools · Adhesives · used in temporary mounting · POSITIONING · ATTACHMENTS

Working posture: Natural standing · Squatting posture · Crawling-about · Work by hand-feel · Crawling-in posture

Working space: Space large · Restrictive · Can't get the tool in · Can't get the hand in · Not visible

Working position: From outside · From inside · Stepping into the car · From under

Identification: Yes for parts · None for parts · Error-proof, yes · Error-proof, N o

Torque control: Common tightening · Torque control · Spot torque control?

Tools: No tool · One touch tool · Standard air tool · Special air · No Air Required

Adhesives: Both-sided tape · In-house tape mounting · Adhesive mounting

POSITIONING: Temporary mounting · No. Temporary mounts · Joint operation · One touch · Hole matching · Butt Positioning · Positioning

ATTACHMENTS: One touch · One touch, more · A few Bolt & Nut · Many Bolt & Nut · Special harnessed parts

Co. A / Ours / Co. B

Factors Causing Differences in Man-hours

Man-hours		Parts or No Parts	No. of Parts (Parts Consolidation)	Diff. In No of Fixings	Diff. in Sems use	No Air Required	Common tightening	Yes for parts	From outside	Space large	Natural standing

Diff. in: Man-hours · Parts · No. of Parts · Diff. In No of Fixings · Diff. in Sems use · No Need of Positioning · Temporary Attaching work · Diff. in Need of Temporary Attaching work · Affected by Adhesives · Diff. in Need of Use Tools · Diff. in Blemishes of Adjustment/Reaction · Diff. in Identify action · Diff. in Torque Control · Diff. in Working Positions

Man-hours: Net · Diff.

Comparison of Man-hours — Co. A / Co. B

Evaluation: Man-hours · Mate-rials · Product Suitability · Over all Evaluation

A-3

Cost Tear Down Report

Date:

Analyzed and reported by:

Item	Part number				
	Parts Name	Weight	Q'ty/unit	Price	
		lb.		Unit price	Per set

Features (major specification differences, sales points, patents, etc.)

Weight

Conclusion	Cost Level		
Co. A	%		lb.
Co. B	%		lb.

Factors causing cost differences	**Co. A** (+ -)	**Co. B** (+ -)
(Company)	(Company)	

Factors for low cost

Factors for high cost

Actions: Current and Future

Target (difference)	
Weight	lb/set
Cost	$ /set

A-4

Cost Tear-Down Analysiss/Tally Sheet (Assembly /Part)

Evaluator: _____ Date: _____

			Our Product					Company A				Company B			Future Model	
No	Components	No	Parts No	Parts Name	Q'ty	Unit Cost	Cost/set	Measured	Points of Interest	Cost/set	Difference	Measured	Points of Interest	Cost/set	Difference	Measured	Points of Interest	Target Cost	Difference
1																

Manufacturing cost

Cost Tear Down: List of Improvement Items

Parts to be anlyzed	Product No.	Analysis Date	Section	Name		
No.		Improvement Description		Invested Amount	Savings Amount	Adopt.or Non-adopt

Material Tear-Down Data Sheet (for Comparison with Our Own Parts to Review Material & Yield rate)

No.	Part Name Part Number	Required Function & Restricting Condition	Report: Product model	Date Output No./Yr.	Cost ($)	Weight (lb.)	Org'n lb Price ($/lb)	Mat'l Used	Name Mat'l Price ($/lb)	Yield	Remarks

Material Tear-Down Data Sheet (for Comparison with Our Own Parts to Review Material & Surface Treatment)

| | | | Report: | Date | | | | Section | | Name | | |
No.	Part Name / Part Number	Required Function & Restricting Condition	Product model	Output No./Yr.	Cost ($)	Weight (lb.)	Unit price ($/lb)	Mat.U.Price ($/lb)	Surface treatment	Heat treatment	Remarks

A-8

Material Tear Down Data Sheet for Comparison with Competitors' Parts

(P191)

No.	Part Name / Part Number	Required Function & Restricting Condition	Report Date / Product model	Org'n / Output No./Yr.	Cost ($)	Weight (lb.)	lb Price ($/lb)	Mat'l Used	Mat'l Price ($/lb)	Name / Surface treatment	Heat treatment	Remarks

A-9

Matrix Tear-Down Report (Assembly)

Item	Part number		No. of types	Total investment	Output Quantity			No. of use per unit		Date		Adaptable car model						Total output – No./month
	Part description				Max	Min	Average		/units	company								
										Reporter								
										Analysis		Output qty. by models						

Analysis	Base item	No.	Market Introduction Yr./Mo. of	Difference from the base	Difference from specifications	Other factors (comments)	Weight (lb.)	Weight change +/-	Cost index (%)	Additional investment
Item No.										
Part No.										
1										
2										
3										
4										
5										
6										
7										
8										
9										
10										

Comments Regarding Future Applications

Remarks

A-10

Process Tear Down Report (Assembly or Part)

Analysis	Date		Org. Unit		Name	

Analysis	Tear-down Object	Base Process	Parts No. Parts Name		No. Used /product unit	Output Volume			Scope of Proces		A S M	Parts No. Only when analyzing part Name	
						Max	Min	Average					

Process No.	Name of Process (No. of Machine Used, No. of Mold/Die/Model Used)	Descrption of the Process			Time in Sec.,			Parts Numbers								Remarks
			Output Volume (units per month)		Manual	Automatic	Walking									
					-	-	-									

Sketch		Bacic Spec	Total	NO. of Process Manhours										
				Differences in Spec.										

Dynamic Tear Down Man-hours Analysis Comparsion

Part | Cover Assy Part Name

Our product | **Competitor**

Competitor

Sketch of Assembly Process

Sketch of Assembly Process

SAMPLE

Total Assembly Time

Total Assembly Time

Process

Structural Elements & Related Assembly Facors

Assembly Factors

BOLT			SCREW			TAP SCREW			NUT			WAS...			CLIP			BAND			PIN		OTHER			

BOLT: Non Washer, Flange, One touch, more, Single Sems, Double Sems
SCREW: Non Washer, Single Sems, Double Sems
TAP SCREW: Non Washer, Single Sems, Double Sems
NUT: Single Flange, Single Sems, Resin, Spire
WAS...: Plain, Spring, Lock
CLIP: Welding, Anchor, Hose, Wire, Plate Band
BAND: Rubber, Resin, Metal
PIN: Split pin, Portable Jack?
OTHER: Sealer, Bond, Tape

Total No. Fixings
Total Assembly Time
Weight

Co. A / Ours / Co. B

Comparison of Work Methods (by Assembly Factors)

Man-hours

ATTACHMENTS	POSITIONING	Adhesives	Tools	Torque control	Identification	Working position	Working space	Working posture	Overall evaluation

ATTACHMENTS: One touch, One touch, more, A few Bolt & Nut, Many Bolt & Nut, Special harnessed parts
POSITIONING: One touch, Hole matching, Butt Positioning, Positioning
(ease of temporary position): No Temporary mounts, Temporary mounting, Joint operation
Adhesives: Both-sided tape, In-house tape mounting, Adhesive mounting
Tools: No tool, One touch tool, Standard air tool, Special air, No Air Required
Torque control: Common tightening, Torque control, Spur torque control?
Identification: Yes for parts, None for parts, Error-proof, yes, Error-proof, N o
Working position: From outside, From inside, Stepping into the car, From under
Working space: Space large, Restrictive, Can't get the tool in, Can't get the hand in, Not visible
Working posture: Natural standing, Squatting posture, Crawling-about, Work by hand-feel, Crawling-in posture

Co. A / Ours / Co. B

REMARKS

Factors Causing Differences in Man-hours

Man-hours		Diff in Sems use	Diff in No of Fixings	No. of Parts temporary mounting	No. of Parts Parts Consolidation	Diff in Man-hours	Diff in Working Postures	Diff in Torque Control	Diff. in Identify-ication	Diff. in amount of Adjustment Revision	Diff in Working Postures	Overall Evaluation

Evaluation

Man-hours	Mate-rials	Product Saleability	Over all Evaluation

A-12

BIBLIOGRAPHY

Proceedings from the 26th National Convention on Standardization." Japan Standards Association (1983), pp. 113-118.

"Standardization and Quality Control", Japan Standards Association: Vol. 37 (1984.2), pp. 35-42.

Kaufman, J. Jerry & Carter, Jimmie L., "Evaluating Brainstorming Ideas," SAVE Proceedings: Vol. XIX (1994), pp. 186-194.

Kaufman, J. Jerry, Value Engineering For The Practitioner, North Carolina State University, Raleigh, NC, (1985).

Kaufman, J. Jerry, "Value Management: A Methodology, Not A Tool", Value World, Vol. 15 No. 1 (1992), p.13-17.

Kaufman, J. Jerry, Value Management, Creating Competitive Advantage, Crisp Publications, (1999).

Sato, Yoshihiko, "Cheer up with VE." Monthly Series (a 4-6 page column) 1997.4-1999.3 Issues, Nikkei Mechanical, Nikkei BP.

Sato, Yoshihiko, Value Engineering, U-Leag Press, (1996).

Sato, Yoshihiko (ed.), "Tear-Down Manual—VE Training Video", Yasui Electric Publishing Co., Tokyo, (1996).

Tanaka, Masayasu, "Cost Planning—Theory and Practice", Chuo Keizai Press, Tokyo, (1995).

ABOUT THE AUTHORS

J. Jerry Kaufman, BS, CVS (Life), FSAVE

Jerry Kaufman, founder of J. J. Kaufman Associates, Inc., a Value Management services company, has over 35 years of Design Engineering, Value Engineering and corporate management experience in the industrial, electronic, processes, services and aerospace markets.

Formerly Corporate Director of Value Engineering for Cooper Industries, a multibillion dollar corporation, his experience includes 25 years of progressively more responsible management positions involving products, processes, and business management. Jerry is a Past President, Certified Value Specialist (CVS) and Past Board Chairman of the Society of American Value Engineers (SAVE), and internationally known and honored in his field. He has been cited in "Engineer's Joint Council in 1973, received the "Meritorious Services Award," and is a "Fellow" and "Life Member" of SAVE International. In 1980 and again in 1999 he was awarded the "Presidential Citation" by the Society for Japanese Value Engineering for his contributions in establishing and supporting Value Engineering in Japan. The American Association of Cost Engineers (AACE) recognized Jerry as their "Speaker of the Year" for the period 1989/1990.

In 1994 SAVE presented The "Lawrence D. Miles Award," its highest SAVE achievement to Jerry for his many contributions to the field of Value Management. Jerry joins a select group of recipients to receive this rarely presented award since its inception 30 years ago.

Jerry has written and presented over 20 papers, four textbooks and contributed a section on Value Engineering in the Encyclopedia of Management, published by Van Nostrand Reinhold. He has also contributed numerous articles for Japanese, European and U.S. publications on advanced Value Management concepts. His third book titled "Value Engineering For The Practitioner," published by North Carolina State University (NCSU) is used to train VE Practitioners. Jerry's most recent book titled "Value Management – Creating Competitive Advantage" explores new VM techniques and management issues. Jerry served as an instructor for NCSU and McGill University's

Management Institute, in Montreal Canada, where he taught a series of industrial extension courses in Value Management.

Jerry has also conducted Value Management seminars in Canada, Europe, England, Japan, and Korea. As President of J. J. Kaufman Associates he has provided consultant and training support services and conducted VM Workshops related to product development, manufacturing, process improvement (energy and chemical), Total Quality Management, and Organization Effectiveness (Business System Re-engineering) for many multi-national corporations around the world.

Home Address;
12006 Indian Wells, Houston, Texas, 77066, USA
Telephone: (281) 444-6887
E-mail: jerry2k@swbell.net

YOSHIHIKO SATO, CVS-Life, FSAVE, J-MCMC

Yoshihiko Sato, Certified Value Specialist, is President of the Value & Profit Management (VPM) Technical Institute Inc. in Sagamihara, Japan, and Associate Director of the Society of Japanese Value Engineering (SJVE). Mr. Sato Joined Isuzu in 1963 and advanced to Senior Scientist, Engineering Operations Office, Isuzu Motors, Japan. In 1963 he began his career with Isuzu, working in the Production Engineering Department. In 1972 while he was assigned to the Cost Planning Department, Mr. Sato developed the "Tear Down" process as a way to implement the value analysis methodology into the operating culture of Isuzu to dramatically increase Isuzu's global competitiveness.

In 1993 Mr. Sato became General Manager of the Isuzu Cost Management Department and advanced to General Manager of Cost Technology Department. In 1988 Mr. Sato was one of the few international members to qualify as a Certified Value Specialist (CVS), as approved by SAVE International.

In 1977, after five years of development work, he presented his VA Tear-Down method at a seminar at Sanno College, Tokyo. Following this presentation, eleven automotive companies and many leading electronic companies in Japan adopted the Tear-Down method.

Mr. Sato was elected a Fellow by the SAVE International Board in 1995. In the same year he received a VE Promotion Award from SJVE. In 1998 he received the VE Outstanding Research and Development Award from SJVE. He was the first member to receive two prestigious awards at an annual SJVE Conference. His J-MCMC status, awarded in 2000, represents a Master Management Consultant certification by the All-Japan Management Consulting Organizations. In 1999 he accepted retirement from Isuzu to form the VMP consultant company. Currently, the sphere of his VA and Tear-down consulting activities extends to clients in Japan, Taiwan, South Korea, and other Pacific rim countries.

Mr. Sato has published many papers and presented lectures on his Value Engineering and Tear-down concepts, and methods, and accepted many speaking invitations throughout the world to present these concepts. Mr. Sato has authored four books and produced two software publications, as listed below:

Promoting In-Company Standards (Japan Standards Association), 1987

Value Engineering (ULEAG Company), 1996

All of the Tear-Down (Nikkei BP Company), 1997
Genki ga Deru VE (The Power of VE) (Nikkei BP Company), 1999
All of the Tear-Down (translated in Chinese in Taiwan) 1999 Video:
Ultimate Tear Down (Yasui Electrical Publisher), 1998
Software: Cost check 3055 (2000) (Nikkei BP Company), 2000

Home Address;
Kanumadai, Sagamihara-shi, Kanagawa-ken, Japan 229-0033
Telephone: +81-(0)42-753-2232   Fax: +81-(0)42-753-8658
E-mail: vpm_y-sato@cam.hi-ho.ne.jp

INDEX